Norton pcANYWHERE™

The Complete Communications Guide

Jack Nimersheim

Windcrest®/McGraw-Hill

FIRST EDITION
FIRST PRINTING

Published by Windcrest Books, an imprint of TAB Books.
TAB Books is a division of McGraw-Hill, Inc.
The name "Windcrest" is a registered trademark of TAB Books.

Library of Congress Cataloging-in-Publication Data

Nimersheim, Jack.
Norton pcAnywhere : the complete communications guide / by Jack Nimersheim.
p. cm.
Includes index.
ISBN 0-8306-4021-5 (hard) ISBN 0-8306-4022-3 (paper)
1. Communications software. 2. Norton pcAnywhere (Computer program) I. Title. II. Title: Norton pcAnywhere.
TK5105.9.N563 1992
005.7′1369—dc20 92-7241
CIP

TAB Books offers software for sale. For information and a catalog, please contact TAB Software Department, Blue Ridge Summit, PA 17294-0850.

Acquisitions Editor: Brad Schepp
Book Editor: Kellie Hagan
Director of Production: Katherine G. Brown
Series Design: Jaclyn J. Boone
Paperbound Cover: Sandra Blair Design
and Brent Blair Photography, Harrisburg, PA WT1

Contents

Acknowledgments

Before getting into the nuts and bolts of using Norton pcANYWHERE, I'd like to take a few moments to thank some special people, without whose assistance I could not have written this book. They include:

Brad Schepp, acquisitions editor for TAB books. When Brad and I first met at COMDEX/Spring, 1990, it didn't take us long to decide that we wanted to work together on a project. You're holding the book that grew out of those initial conversations in your hands.

Virginia Bare, Nancy Stevenson, Jackie Brinker, Michael S. Kerman, and Terri Sammonds at Symantec Corporation, the company responsible for pcANYWHERE. When I needed a piece of inside information or a question answered, one or more of these kind individuals responded to my request. It's safe to say that, without them, this book would never have been completed.

Kellie Hagan, the TAB editor whose deft touch never failed to improve my original draft. (Writers love subtle editors; Kellie fits that description to a T.)

Ralph Roberts, friend and fellow writer, for his clear and concise explanation of Internet, which you'll find in appendix B. (If I didn't like Ralph so darn much, it would be easy to begrudge him his talent. He could describe quantum physics using language anyone could understand.)

Susan and Jason, my wife and five-year-old son, simply for being who they are. I couldn't accomplish anything without these two—and even if I did, it wouldn't mean as much. They have my love, always.

Foreword

Warning! Telecommunications can be hazardous to the way you think about using your computer. Just when you thought you had mastered your computer, somebody like Jack Nimersheim has to come along and open a whole new cosmos to you. Because he might not get around to it, let me say that serious addiction to this new environment is likely.

Reading Jack's work is more like reading your favorite type of fiction. It's easy, informative and when you finish you can say "I understand!" Jack has put himself in your shoes and written down what you need to know in a way that makes it easy to understand. It's clear, concise text that's easy to read with illustrations to enhance your ability to become an effective telecommunicator.

Once you've experienced the thrill of connecting to the ever-increasing universe of electronic information exchange, your life will change. From local bulletin board systems (BBSs) through subscription information distribution services (Delphi, CompuServe, BIX, etc.), the wealth of information along with the host of people available is beyond comprehension. If you look long enough and hard enough, you can find someone to communicate with anywhere that has power and phone service. For those of you who are into amateur radio, your universe just got infinitely larger too.

By now you might have asked yourself "who is Dick Evans?" and "why should I believe him?" My username on Delphi is ELLISCO. If you look for Jack, his username is NIMS. Ralph Roberts, who you'll meet later in this book, uses AUTHOR. I've been a citizen of the electronic universe since 1984 and currently serve as a Systems Assistant for Delphi. Quickly defined, this means I'm available to help users and information providers get the most out of their time and effort while they're on-line.

Having said all of this, I'd better summarize it quickly before my time and space expires. Relax and read this book through once to get the feel of Norton pcANYWHERE and the vastness of the electronic universe that's accessible when you use it. Then start making friends locally. From there your vistas will expand until they're unlimited. Enjoy.

Dick Evans
(aka ELLISCO)
Systems Assistant
Delphi, a service of General Videotex Corp.
Cambridge, MA

Introduction

Fasten your psychic seat belts, folks. We're about to embark on a journey around the world. Even if you're afraid of flying, as many people are, don't worry. This particular trip can be made without ever leaving the comfort of your own home.

How is this possible? Through a miracle of our modern PC age called telecommunications. I don't want to get too technical this early in the game, so I'll simply state that telecommunications involves placing two computers in contact with one another over standard telephone lines.

Telecommunications allows a PC in Miami to call a computer in Madrid. It lets someone in Cuba contact CompuServe, a commercial information service located in Columbus, Ohio. (*Commercial information service* is another PC-related concept I'll explore further a little later in this book.) Telecommunications, to introduce a phrase you'll encounter several times in the following pages, places the world at your fingertips—hence, my earlier reference to the scope of our travels together.

Your flight attendant for this electronic excursion will be Norton pcANYWHERE, a program I first discovered several years ago, when I was looking for a practical way to contact and take control of several computers that were separated from my own system by several miles. Back then, I was a computer consultant specializing in programming database applications for a number of small businesses located in and around Cincinnati, Ohio. On several occasions, pcANYWHERE helped me solve problems that otherwise would have required an on-site visit to one of my clients.

Both pcANYWHERE and I have changed a lot since then. For a while, we each went our separate ways. I gave up computer consulting to become a full-time writer, and pcANYWHERE slowly evolved into a powerful program that can be used for a wide range of activities beyond the remote computing for which it was originally designed. It was also purchased by Symantec Corporation and placed in the care of that company's Norton Products division. Hence, the subtle but obvious name change. For simplicity's sake, I'll refer to it by the more generic *pcANYWHERE* throughout this book.

Thanks to the good folks at TAB Books, pcANYWHERE and I are back together again. It's a reunion I've enjoyed thoroughly.

Jack Nimersheim
Vevay, Indiana
January, 1992

1
CHAPTER

What is Norton pcANYWHERE?

Have you ever wished you could be in two places at the same time? You might not realize it, but, to a certain degree, you can. This is what occurs, in a manner of speaking, every time you use a telephone.

Think about it. All you have to do is pick up the telephone receiver, punch in a few numbers, and you can visit Paris or Peru, Bangkok or Burma, Madrid or Moscow—or merely drop in on a friend down the street—without ever once leaving the comfort of your own home. Calling distant places on the telephone has become so much a part of our daily lives that we tend to forget how significant a technological advance over previous forms of communication it is. (Imagine having to stoke the fires and generate smoke signals every time you felt like chatting with a neighbor in the next town.)

Now consider for a moment another relatively recent miracle of modern technology—the personal computer, or PC. Whereas a telephone's primary function is to simplify the process of interacting with others, a personal computer, as its name implies, excels at augmenting our personal skills. Today, several decades into the so-called "computer revolution," only a true technophobe would try to deny that a personal computer can perform a wide range of activities faster, and with far greater accuracy, than the person using it.

Imagine what you might accomplish, therefore, if there were some way to take these two amazing technologies, the telephone and the personal computer, and combine them into a single productivity tool. Were this possible, a PC could travel almost anywhere, at almost any time. A PC? Anywhere? pcANYWHERE. A pretty impressive idea, and the obvious name for a program that allows you to accomplish exactly that.

pcANYWHERE places the world, quite literally, at your fingertips. By integrating computer and communications technology, a process called *telecommunications*, pcANYWHERE allows you to electronically traverse the globe. With pcANYWHERE installed on your PC, you can communicate with (and even take control of) another computer over a standard telephone line.

I'll discuss specific aspects of telecommunications in the next chapter; understanding what telecommunications is and what it involves will help you use pcANYWHERE more effectively. Before getting into a lot of technical stuff, however, it's probably a good idea to examine, in a more general way, just how pcANYWHERE can be used—that is, some of the different types of tasks it allows you to perform.

Anywhere, anytime

The key to pcANYWHERE's potential lies in the two words that make up the title of this section: anywhere and anytime. Like any good telecommunications program, pcANYWHERE shatters the twin barriers of space and time. pcANYWHERE eliminates logistical problems commonly associated with distance because, as I mentioned earlier, today's global communications network places cities on the opposite side of the globe a mere phone call away. Furthermore, pcANYWHERE virtually eliminates time constraints. That's because, unlike using traditional distribution channels—rail transportation, trucking, air freight, even the U.S. Postal Service—moving information (the primary product of the computer age) over a telecommunications link is a near instantaneous proposition. Let me give you one quick example of what I mean.

Destination: Knoxville

A few months ago, we received an invitation to my wife's family reunion, an annual event that's held in a different part of the country each year. This year, the festivities were in Knoxville, Tennessee, close enough to our Indiana home that we could plan to attend, which we did.

The location might have been convenient, but the timing, as it turned out, was terrible. I was facing tight deadlines for a number of writing assignments as the weekend of the reunion approached. Each of these required that I be able to pass information quickly back and forth with the editors of several different computer magazines. To make matters worse, one particular assignment could not be completed without first testing a program that, because of its extensive hardware requirements, would not run on the notebook computer I'd originally planned to take to Susan's family reunion with me. Unless I could figure out some way to be in two places at once, therefore, things were looking pretty gloomy for the the Nimersheim clan, relative to the mini-vacation we'd planned to squeeze out of our trip to Knoxville and, schedule permitting, the mountains beyond.

Enter, pcANYWHERE

I keep coming back to that two-places-at-once theme, don't I? This is no accident. pcANYWHERE's primary function, you see, is to let you use one computer to call up and take control of a second computer in a different location. This process, illustrated in Fig. 1-1, is commonly referred to as *remote computing*.

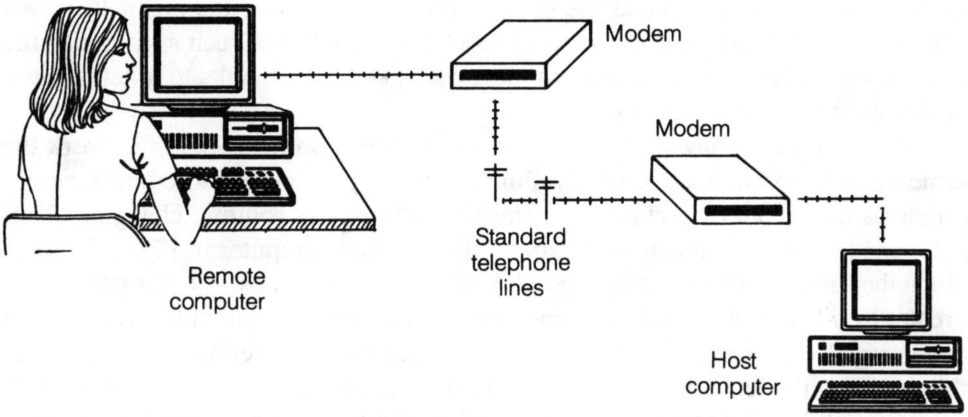

1-1 pcANYWHERE's remote-computing features allow you to be two places at once.

You might want to examine this figure for a moment. It contains some critical terms and concepts you'll encounter regularly in the chapters ahead.

Notice, for example, that the PC on the left-hand side of the illustration is identified as a *remote computer*. The person sitting at the keyboard of this PC might look like he's working on his own system. In truth, however, he's running a program installed on the PC located to the right of the figure, identified as the *host computer*. Taking a few moments now to understand the respective role each of these two systems plays in a remote-computing session will greatly simplify the task before you—that is, learning how to set up and use one specific remote-computing program, pcANYWHERE.

A gracious host

If you think about it, the computer to the right of Fig. 1-1 is not that different from the host of a party. It has, in essence, opened an electronic door (the telephone connection) and extended an invitation to the remote system to stop by for a visit, as it were. And make no mistake about it—the host computer in a remote-computing session is a gracious host, indeed. It actually turns control of its most valuable resources—the hardware and software that transform any PC into something more than an overpriced paperweight—over to the person using the remote computer.

To bring this discussion back to the subject of the book, the host computer in

Fig. 1-1 is the system on which you will later install and run a special program called AWHOST, which is included in the pcANYWHERE package.

PC couch potato

To understand the role of the remote computer in a pcANYWHERE connection, you need look no further than a contemporary television set or VCR. I don't know about you, but I grew up in the pre-couch potato era, when changing television channels actually required that you get up from a chair or sofa and manually turn a knob to the desired station. (Hard to believe, isn't it?) Today I can accomplish the same thing without budging from my favorite easy chair, thanks to a handy little device called a remote control.

These modern conveniences—no two of which, I sometimes think, work the same or incorporate even remotely similar designs—allow you to control certain functions on a modern television set (raising or lowering volume, changing channels, and the like) from across the room. The remote computer in Fig. 1-1 serves much the same function during a pcANYWHERE session, except that its range is greater by a factor of several thousand. It can control a host computer from almost anywhere in the world, so long as it has easy access to a telephone line. Later in this book you'll learn how to install and use AWREMOTE, pcANYWHERE's remote module, so that your PC can function like the remote computer shown in Fig. 1-1.

Putting it all together

Having laid this critical groundwork, let me return to the family reunion anecdote, which I left unresolved a few paragraphs back. The story has a happy ending. By installing the appropriate pcANYWHERE module on both my desktop PC and the notebook I carried to Knoxville, I was able to dial up home from our hotel room, test that hardware-hungry program I had agreed to review, complete all my assignments on deadline, and still have time to accompany my five-year-old son on his first trip up and over the Great Smokies.

Stated simply, pcANYWHERE's remote-computing capabilities allowed me to carry my office with me. This, in turn, let me continue working, despite the 250 miles separating me from my normal homebase. Consequently, and even more importantly, I wasn't forced to cancel the mini-vacation I had promised my family.

Remote possibilities

My admittedly personal tale only scratches the surface, however, when it comes to analyzing the potential of this amazing program. Whatever your chosen profession, if it involves PCs, chances are it can benefit from the addition of pcANY-

WHERE to your software arsenal. Other possible uses for pcANYWHERE include:

Project coordination pcANYWHERE's remote-computing capabilities can allow coworkers in multiple sites to "look over each other's shoulders," as it were, while they develop different phases of a complex project.

Sales A pcANYWHERE host system can serve as a hub for a centralized ordering system. Setting up procedures that allow your sales force to call in customer orders quickly could decrease the turnaround time otherwise required to pull inventory and ship products.

Off-site technical support The ability to remotely run a client's PC can allow you to recreate and analyze any problems he might be having with a particular piece of hardware or software. Depending on the situation, it might even be possible to resolve these problems without having to schedule an on-site visit. You could, for example, use pcANYWHERE to reconfigure a troublesome piece of software over a remote connection.

Training Imagine being able to remotely walk someone through the precise steps required to use a new application program correctly. For complicated operations, pcANYWHERE's "chat" feature can be used to explain to the person on the other end of the remote connection exactly what he should be doing, each step of the way.

As these few scenarios demonstrate, pcANYWHERE's remote-computing capabilities can be used to improve the efficiency of a wide range of PC-related activities. But remote computing is only one aspect of pcANYWHERE—albeit, the most publicized one. pcANYWHERE can be used to perform other types of on-line activities, as well.

Additional on-line activities

Running that troublesome program remotely was only the most obvious task pcANYWHERE allowed me to accomplish in Knoxville. I also relied on pcANY WHERE to keep in touch with several editors from my hotel room by using a variety of electronic messaging services. I even called on pcANYWHERE to help me electronically transmit my finished assignments, thus avoiding every writer's worst nightmare: a missed deadline.

Had pcANYWHERE supported only remote computing, I wouldn't have been able to use it for all my Knoxville activities. Instead, I would have had to rely on a second, more traditional communications package such as Procomm Plus or Crosstalk to exchange messages and files with my editors. But why carry along multiple programs when one will suffice? For many people, the answer to this question might lie in the fact that the pcANYWHERE documentation, which does an excellent job of explaining the program's remote-computing features, does not provide much information about additional on-line activities. A lot of pcANYWHERE users simply don't recognize the program's full potential. That's a situation I hope to resolve in this book.

As the title proclaims, this is pcANYWHERE: The *Complete* Communications Guide. Let me reiterate a statement I made earlier. pcANYWHERE puts the entire world, quite literally, at your fingertips. Once you know how to use pcANYWHERE, the list of activities beyond remote computing that it supports is as diverse—and, in some cases, equally exotic—as the places you can call.

In addition to remote computing, other potential uses for pcANYWHERE include:

- Electronic messaging
- Real-time conferencing
- File transfers
- On-line research
- Electronic transactions
- Information gathering

Throughout the chapters that follow, you'll be examining how pcANYWHERE allows you to perform these various on-line activities. For now, however, lets take a brief look at what each entails.

Electronic messaging

Electronic messaging, or e-mail, has taken the world by storm. Recent studies suggest that anywhere from 8 to 15 million people rely on electronic messaging for at least a portion of their correspondence.

The principle behind electronic messaging is simple: You forward an electronic message to someone else (usually to an electronic "mailbox" on a commercial information service or dedicated e-mail service) and, if appropriate, they respond. This response, however, inevitably takes some time. For this reason, electronic messaging, like traditional mail, is considered a *time-shifted* activity.

There are two primary advantages that electronic messaging possesses over the much-maligned U.S. Postal Service: speed and convenience. Electronic messages can be passed back and forth, if not instantaneously, then at least in a matter of seconds, regardless of the distances involved. For example, during my recent trip to Knoxville, it took about two minutes for me to compose and send a letter to an editor in California via his electronic mailbox, which just happened to be located in Washington, D.C. If the intended recipient of a given message has been alerted to its arrival, he can dispatch a response with equal speed.

Electronic messaging is also convenient, especially for those on the receiving end of things. People who subscribe to an electronic mail service no longer need to wait days for the mail carrier to deliver important messages. Rather, they can check on its status as their schedule permits by simply dialing up the appropriate phone number and seeing whether any messages have been delivered to their electronic mailbox. (One caveat: with commercial information services like CompuServe or Delphi, you'll be charged a nominal fee each time you sign on to check your electronic mailbox. Even small charges can add up to a rather substantial monthly bill when it comes to telecommunicating. I'll discuss the costs of going on-line in the next chapter, *A telecommunications primer*.)

Electronic messaging promises to revolutionize the manner in which we communicate with one another. To share in the advantages associated with this new communication medium, however, you will first need to get pcANYWHERE up and running on your PC. Don't worry. I'll begin outlining the steps involved in accomplishing this shortly. For now, you still have a few more on-line activities to examine.

Real-time conferencing

If electronic messaging resembles the U.S Postal Service, the best analogy I can come up with for real-time conferencing is that of using a multi-line telephone to participate in a conference call—or, perhaps even more appropriate, a computerized variation on the CB craze that swept the nation a few years back. Callers participating in a real-time conference "chat" with one another by typing in comments at their keyboards. Pressing the Enter key causes what you have typed to be displayed on the screens of everyone else involved in the current conference.

Almost all commercial information services include some form of real-time conferencing among their offerings. For example, every Thursday night I try to stop by the Writers' Conference on Delphi, one popular on-line service. That's when established and would-be wordsmiths get together to discuss topics relating to the so-called "writing life." Some weeks are devoted to structured discussions. More common, however, are informal gatherings, where we all just show up and let the conversations proceed at their own pace. With a dozen or more people participating in this conference on any given Thursday night, something of interest almost always comes up. Like most conference groups, ours evolved informally over time, as I and several of my fellow Delphi users realized that we shared a common interest.

File transfers

One of the more practical uses for pcANYWHERE is to exchange files between two computers. To some extent, this is what happens when you send someone an electronic message. But the information you transfer with pcANYWHERE need not be limited to simple correspondence. Even more useful is the ability to transmit actual program and data files across phone lines.

Suppose, for example, that your supervisor in corporate headquarters has decided that she wants to see, ASAP, next year's budget for your local operation. Using pcANYWHERE, you could transmit the actual file containing the assumptions and calculations used to create your projections, for her review. Viola! Instant communications and, quite possibly, a glowing evaluation from the boss based on your efficiency.

On-line research

How would you like to be able to access a wealth of information literally at the touch of a button—or, in this case, a keyboard? That's about the best description I can come up with for on-line research.

If you need a fact or figure, whether it be of a general or specialized nature, chances are that this information exists someplace, in some form, on some on-line database, waiting for you to find it. News services like the Associated Press and Reuters, financial information from sources as diverse as the New York Stock Exchange and Dow/Jones, historical records, business news, economic studies, medical research, and statistics galore—each of these areas of interest is covered, in one form or another, on a number of public and private on-line services. Most commercial services also offer easy access to one or more on-line encyclopedias that can be used to research more general projects such as homework assignments, term papers, and the like.

Be aware, however, that most research databases carry with them steep access charges—that is, the price you'll pay for any time you spend on-line using them. The per-minute rate for using a specialized research database is generally higher than that charged by the more popular commercial services such as CompuServe and GEnie. Some commercial services also levy an additional surcharge, on top of their regular rates, for any time you spend accessing a dedicated research database such as an on-line encyclopedia or wire service.

Electronic transactions

Once upon a time, a very young J. Nimersheim read a science fiction story about a man who lived his entire life in a single room. The main premise behind this story (the name and author of which, unfortunately, have been wiped from my mind's slate by the intervening years) was that technology had advanced to such a degree that everything a person might ever want or need could be either procured by or simulated with a powerful "thinking machine," if I recall the terminology used. This thinking machine—at the time, nothing more than a product of a writer's creative imagination—bore a close resemblance to today's computers.

The protagonist in this story never left the safety of his bountiful yet artificial environment. Why should he? Everything he needed was there. (To give you some idea of how imaginative this unknown author was, consider how closely his plot, which I'm sure seemed far-fetched at the time, resembles the relatively new branch of computer sciences that deals with *virtual reality*.)

Although our own society is not yet this machine-dependent—and, in truth, I hope it never will be—we're definitely beginning to rely more and more on gadgets and gizmos to simplify our lives. It's now possible, for example, to order a wide variety of goods and services on-line through either dedicated computer stores or commercial information services. Financial transactions such as buying and selling stocks, electronic banking, and even automating your monthly check-writing activities, can also be initiated from the comfort of your own home using a communications program like pcANYWHERE.

Will there come a time when, like the main character in that dimly remembered story, we'll be able to fulfill all our needs without ever once stepping out the front door? As I stated earlier, I certainly hope not. Used wisely, however, the ability to shop at home can be a welcome convenience in these modern times when, more often than not, time is one of our most valuable commodities.

Information gathering

Imagine that you could subscribe to a newspaper that covered only stories that were important to you. What if, for example, each morning when you went to the door, you found a daily paper devoted exclusively not only to national news, or sports, or business news, or whatever, but also only to topics in which you were deeply interested? The "My Daily News," so to speak.

Several on-line services now allow you to specify search parameters for news stories, financial coverage, and the like so that any stories relating to topics you find interesting will automatically be flagged for your review the next time you call in. What I'm talking about here is exactly the kind of personalized communications "newspaper" described in the previous paragraph—without the paper. (Wow! No more hauling ten pounds of discarded newsprint out to the garbage every week!) On-line services are the perfect vehicle for the ultimate attainment of information gathering process.

Some people would argue that I shouldn't differentiate between research and information gathering. In my mind, however, research generally encompasses a one-shot, project-oriented effort—for example, accessing an on-line database to look up information for a specific report or assignment. The broader category I call information gathering, on the other hand, is an on-going activity, an electronic extension of more traditional research methods—one unique to the computer age, when even such a seemingly mundane activity as reading the news can be personalized to such a degree that the electronic newspaper constructed on my PC will be totally different than the one delivered to my next-door neighbor.

pcANYWHERE system requirements

As the previous examples illustrate, pcANYWHERE is a powerful program. To repeat an already used metaphor, pcANYWHERE provides you with an electronic passport, good for virtually anywhere in the world. You might be surprised, therefore, at how economical this potential global adventure can be. Unlike many other PC applications, telecommunications is a relatively inexpensive proposition, in relation to the resources it demands. The minimum system requirements for pcANYWHERE include:

- an IBM PC/XT, AT, PS/2 or compatible computer
- at least one serial port
- 320K of RAM
- MS-DOS 2.0 or higher
- a hard disk
- a 5.25" or 3.5" floppy disk drive
- a modem

A pretty basic setup, to be sure. Of course, the better equipped your PC, the more options you'll have available to you once you begin using pcANYWHERE. Additional equipment that can enhance your pcANYWHERE operations includes:

- a VGA display
- a printer
- a mouse
- a memory manager such as Quarterdeck's QEMM-386, QRAM, or Qualitas' 386MAX for loading pcANYWHERE into high memory (Beginning with MS-DOS 5, Microsoft has added its own memory manager to this popular operating system. Consequently, you won't need a third-party alternative if your system is running under DOS 5.)

Which and how many of these additional items you want to incorporate into your pcANYWHERE environment depends on the types of remote operations you plan to perform. The most critical thing to keep in mind is that pcANYWHERE demands very little in the way of system resources, especially on the remote side of an on-line connection.

Note: One important way in which a VGA display and a mouse enhances your remote operations is by allowing you to use pcANYWHERE to run Windows applications installed on a host system. Given the amazing success of Windows 3.0, this capability, one of the major upgrades incorporated into the recently released pcANYWHERE 4.5, is a welcome feature, indeed, and one I'll be covering in detail throughout subsequent chapters and exercises in this book.

Reach out and touch

There's one additional item that's absolutely necessary if you expect to use pcANYWHERE: a phone line. This might seem obvious—so obvious, in fact, that you could argue there isn't much need to discuss this critical component of your pcANYWHERE environment. You either already have a telephone line available, or you call the phone company and get one installed. It's that simple, right?

For the most part, yes. Still, there are a couple of items related to using a telephone line for telecommunications, rather than voice communications exclusively, that you need to be aware of in order to help you avoid problems down the road.

Cancel call waiting

First and foremost, think twice about ordering the call-waiting feature offered by many phone companies. Although convenient for voice communications, call waiting can wreak havoc on a pcANYWHERE session. That innocent "beep" that normally heralds an incoming call can destroy a PC-to-PC connection. Sometimes it merely inserts superfluous characters into a transmission. Quite often, though, a call-waiting signal can interrupt an on-line session, completely dropping the connection.

Before giving up totally on call waiting, however, check with your local phone company. Depending on the type of equipment it uses, there might be a way to temporarily eliminate call waiting before you use pcANYWHERE, thus avoiding the problems associated with incoming calls.

To test whether this is true for your phone system, pick up the phone and press the pound sign (#) followed by 70. If you receive a second dial tone, then your phone company uses newer equipment that allows you to temporarily suspend call waiting. (Should you place a call now, anyone attempting to reach you during that call will receive a busy signal. Call waiting goes back into effect automatically for subsequent calls when you hang up the phone.)

Better still, order a second line

Finances permitting, however, you might want to consider installing a second, dedicated phone line for your pcANYWHERE activities. In addition to eliminating any additional problems, if call waiting is installed on your main line, it's simply more convenient to have a second phone line available for your on-line activities. Should troubles arise during a telecommunications session, for example, a two-line setup allows you to call someone with access to the other PC and, working with them, resolve the problem.

Install a line protector

Your PC equipment represents a sizable financial investment. It only makes sense, therefore, to protect your system against accidental damage. The biggest threat to any computer hardware is something called a power surge, a sudden and sporadic change in the electrical current flowing from a wall outlet or even across a telephone line.

Many multi-outlet power strips include built-in surge suppressor circuitry, which protects the equipment plugged into them from being damaged by power fluctuations in your electrical line. Installing one of these relatively inexpensive devices with your PC system can help you avoid a costly repair cost somewhere down the line (pun intended).

You use a standard telephone cord (called an RJ-11 cord) to connect your modem to a phone jack or wall plug, much as you would a typical modular telephone. When installing your modem, you might want to protect it with a telephone-line surge protector, as shown in Fig. 1-2. Doing so guards against any voltage spikes or power surges originating in the telephone lines.

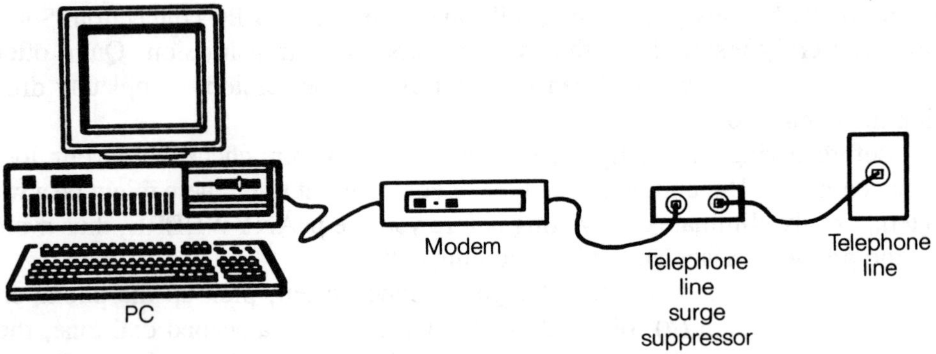

1-2 A line-surge protector guards your system against damage from voltage spikes and power surges.

Tip: Even with a power-line surge suppressor and telephone-line protector installed on your PC, it's a good idea to unplug both your computer system and the RJ-11 cord connecting its modem to the wall plug whenever a thunderstorm strikes. Though useful for controlling common fluctuations in power and telephone lines, no man-made protector can stand up to the unbridled fury of Mother Nature.

Summary

This, in a nutshell, is what pcANYWHERE is, what it does, and what it needs. A lot of information and effort still lies ahead, starting with learning more about the general subject of telecommunications, a topic I'll take up in the next chapter.

2

CHAPTER

A telecommunications primer

My main goal in the previous chapter was to provide a general introduction to pcANYWHERE. In the process I also examined, in a very cursory manner, some of the different types of on-line tasks you'll be able to perform using this amazingly flexible program. I realize you're probably impatient to actually start using pcANYWHERE, to begin traveling this global community I've been talking about. Before doing so, however, we need to take a short detour and discuss some basic concepts relating to telecommunications, that fusion of PC and telephone technology introduced in chapter 1. So let's get started, shall we?

What is telecommunications?

As is true of many PC-related activities, telecommunications carries with it an elusive quality, a certain mystique. What is telecommunications? How does this enigmatic process work? What will you need in order to telecommunicate effectively?

To understand what telecommunications is, you need only take this word and break it down into its component parts. Specifically, Telecommunications is the process of using your PC, a modem, and special software to communicate with another computer over a standard telephone line. *Tele*phone *communications* between two computers. *Telecommunications*.

Telecommunications is a relatively straightforward proposition. Unfortunately, and as anyone who uses a personal computer with any regularity will readily admit, life in PC land is rarely as simple as it first appears.

Like so many other aspects of modern life, the basic concept of telecommunications has become entangled in a web of arcane language and potentially aggravating procedures. Bits, bytes, and baud rate; user IDs and passwords; modems and e-mail; asynchronous communications; protocols and parity. These are but a few of the esoteric terms you'll encounter as you dip your toes into the admittedly murky waters of telecommunications. And make no mistake about it, you'll need to have some idea of what each of these terms means before you can begin using pcANYWHERE efficiently. Having laid this little bombshell on you, let me now introduce a theme that will resurface regularly throughout this book. Namely, *don't panic*. Telecommunicating is not that different from any other PC application. And pcANYWHERE is not unlike other PC programs.

Anyone familiar with Lotus 1-2-3, for example, had to learn about cell references, menu bars, recursive calculations, shortcut keys, @ functions, and a dozen other initially strange concepts before he could feel comfortable using this popular spreadsheet program. Mastering the fine art of "number crunching" requires time, energy, and no small amount of effort. After a while, however, confidence replaces confusion as terms and procedures unique to using an electronic spreadsheet are applied to everyday situations.

The same thing will happen with pcANYWHERE. Once you master a few critical concepts, everything else will fall into place. A good place to begin is with the single most important piece of computer equipment required to use pcANYWHERE—a modem.

The marvelous modem

Stated simply, a modem is a translation device that allows your PC to communicate with other computers, over a standard telephone line. Describing what a modem does and how a modem goes about performing its appointed task, however, requires a slightly more detailed explanation.

What a modem does

If PCs and people spoke the same language, you wouldn't need a modem to telecommunicate. Unfortunately, this isn't the case. That PC sitting on your desk is actually an intricate device containing thousands of tiny electronic switches. Like a light switch, another switching device with which you're no doubt familiar, the electronic switches inside a PC can be set to one of two states: on or off. These are your only options. Furthermore, in the case of these electronic switches, there's not even a smooth transition between the two discrete states. One instant a switch is off, and the next it's on.

Symbolically, the current state of a given electronic switch within a PC is represented by two numerical digits, 0 and 1, identifying whether that switch is off or on, respectively. For this reason, although it's somewhat of a misnomer, computers are commonly referred to as *digital devices*.

By contrast, we—that is, the people who buy and use personal computers—tend to experience events along a continuum. For us, life progresses smoothly, from one moment to the next. If you're out riding your bicycle, for example, you won't suddenly find yourself ten miles down the road from the point at which you started. We lead an existence of smooth transitions from one state to the next, in other words, in an *analog* world.

A modem functions much like an interpreter at the United Nations General Assembly, in that it translates information from one language into another. Specifically, a modem converts the digital signals generated within a PC into analog tones, which can then be transmitted across standard telephone lines, a process illustrated in Fig. 2-1.

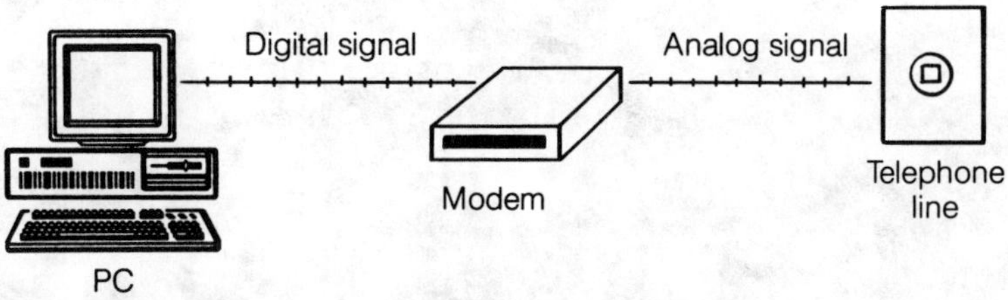

2-1 A modem converts the digital signals from a PC into analog tones, which can then be transmitted across standard phone lines.

How a modem works

The word *modem* is actually a contraction for *mo*dulate/*dem*odulate, the technical terms used to describe how a modem works its electronic magic. As your PC converses with another computer, a modem fulfills its role of electronic translator by performing two critical operations:

- When you transmit information, a modem converts (or modulates) the digital signals your PC generates into analog tones compatible with today's telephone systems.
- When you're on the receiving end of this conversation, the modem converts (or demodulates) the analog tones transmitted from the other computer—or, more correctly, that computer's modem—back into digital signals, which can then be used by your PC.

A modem's ability to perform this digital-to-analog dance is what allows your PC to communicate successfully with other computers over a standard telephone line. Beyond sharing this important similarity, however, all modems are not created equal, if I may borrow an immortal phrase penned by our Founding Fathers. Like the people who use them, modems come in all shapes and sizes—or *configurations*. There are several factors you should consider when deciding which type of modem is right for you, starting with how you plan to connect this critical component of your pcANYWHERE environment to your PC.

The "ins and outs" of selecting a modem

Modems come in two basic models, internal and external. As its name implies, you install an internal modem, also referred to as a modem card (see Fig. 2-2), by inserting it into an open expansion slot inside your PC. Expansion slots, which are generally located along the back of a PC's motherboard, can hold a variety of specialized expansion cards: extra memory, a disk controller, your display card, and the like.

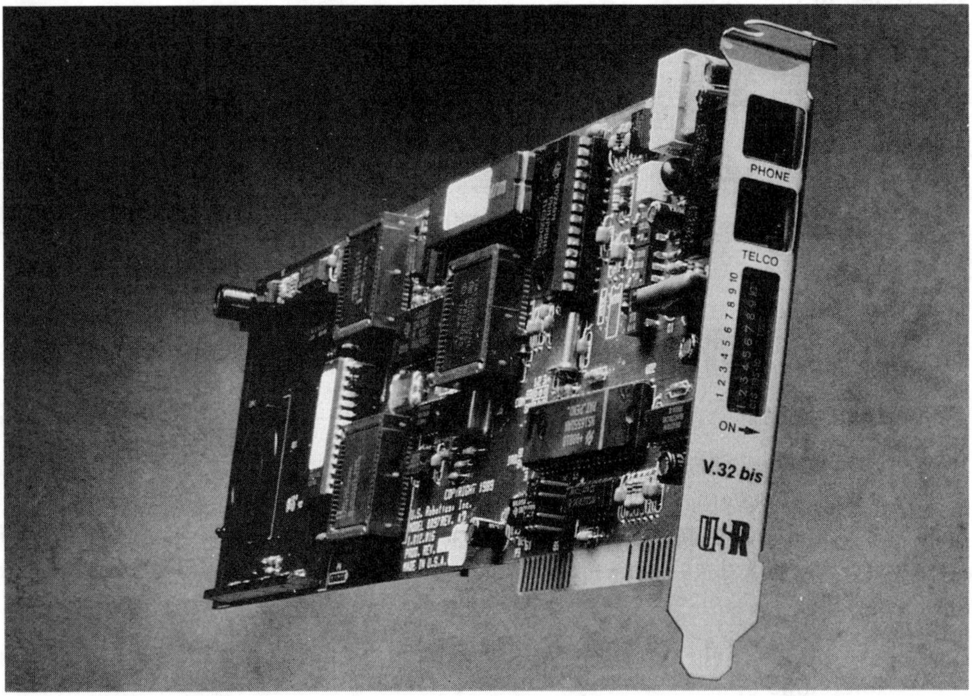

2-2 An internal modem. USRobotics

An external modem, on the other hand, is a separate, stand-alone piece of equipment, similar to the one shown in Fig. 2-3. Because it does not plug directly into the motherboard, an external modem requires a serial port to communicate with your PC. (Don't worry, I haven't talked about serial ports yet. That discussion is coming up in chapter 3, where you'll begin preparing your system to use pcANYWHERE.)

2-3 An external modem. USRobotics

Functionally, internal and external modems are almost identical to one another. Choosing which model is better suited to your situation, therefore, is largely a matter of preference. This decision, however, can be influenced by a number of factors.

For anyone operating on a tight budget, an internal modem is probably the more logical choice. Because they don't require a protective case, internal modems tend to cost less to manufacture (and, therefore, are priced lower) than their external counterparts. At the low end of the modem speed spectrum, you should be able to buy an internal card for about half of what you might expect to pay for a comparable external model.

Of course, if you've already filled all of your computer expansion slots—not an uncommon condition, given the recent popularity of so-called "small footprint" PCs that include only two or three expansion slots on their motherboard—you'll have no choice but to use an external modem. Conversely, if all the serial ports on your PC are being used by other hardware components (say a mouse and a serial printer), then the more logical option is to look for an internal modem.

One note of caution: If you're installing an internal modem in an older system—for example, a first-generation IBM PC/XT—make certain that its power supply is strong enough to support this additional card. Unlike external modems, which plug into a wall outlet, modem cards draw their power directly from the PC's power supply through the expansion slot. Back in the early days of the PC revolution, most manufacturers (including IBM) naively believed that an 85- or 100-W power supply would provide enough get-up-and-go to support any and all additional equipment we PC pioneers might choose to add to our seminal systems. We proved to be a gluttonous bunch, however, and quickly outgrew these conservative predictions. Newer PCs come equipped with more robust power supplies, usually somewhere within the 200- to 250-W range.

A guide to "modemese"

Once you've made the choice between an internal or external modem, the next logical step is to determine an optimum speed, or *data transmission rate*, for your on-line activities. Any consideration of data transmission rates necessarily leads to a discussion of one of the more obscure aspects of telecommunications, the old baud vs. bps dispute. Let me see if I can clear up some of the confusion generally associated with precisely what these terms mean, and how each is used within the broader topic of telecommunications. (We're about to get into some pretty technical stuff, so stay with me.)

Bits, bytes, baud, and other esoteric terms

As pointed out earlier, your PC is actually a complex device containing tens of thousands of tiny electronic switches. Any of these switches, in turn, can be set to one of two states, off or on, with each discrete setting represented by the numeric values 0 or 1, respectively.

For example, the current settings for a series of switches configured as off-on-off-off-off-off-off-on could be depicted as 01000001. Reverting once again to that strange and exotic language called computerese, the appropriate term for each value within this symbolic listing is a *binary digit*, or *bit*. Taken as a whole, therefore, the above notation would reflect the current on/off settings of eight individual bits within your PC.

I must confess, these eight digits look like a lot of gibberish to me; they probably do to you, also. Using something called *ASCII character codes*, however, a computer can take such gibberish and transform it into useful information. (If you're curious, ASCII, pronounced *ass-key*, is an acronym for American Standard Code for Information Interchange.)

ASCII code is similar to the older and more familiar Morse code, in that it uses a series of predefined sequences to represent individual letters, numbers, punctuation marks, and other common symbols. Whereas Morse code relies on unique combinations of dots and dashes, however, ASCII code is built around organized groupings of binary digits, called *bytes*. For example, the sequence shown above, 01000001, is an eight-bit byte that corresponds to the ASCII code for an uppercase *A*. If your PC were to receive this sequence while conversing with another computer over a modem connection, therefore, it would display the letter *A* on your monitor.

Admittedly, this cursory explanation somewhat oversimplifies matters. It does, however, provide a rough idea of how computers telecommunicate—that is, exchange information over a modem connection. From here, it's just a short step to a practical understanding of several familiar but potentially confusing terms often used to measure how quickly data is being transferred between the two systems involved in an on-line session.

Bit rate

Bit rate indicates the speed at which two computers are exchanging data. Bit rate means exactly what you might expect, i.e., the number of individual bits transmitted over the phone line within a specified time period during a given on-line session.

bps

Bit rate is almost always measured in bits per second, or *bps*. Modems are often grouped together based on their optimal bps rating. For example, a 9600-bps modem is one that's capable of transmitting or receiving up to 9600 bits of data every second.

Baud

This seemingly innocent word has caused more confusion down through the years than any other term or concept you're likely to encounter in a discussion about telecommunications. Many people mistakenly believe that baud is synonymous with bit rate or bps. It isn't. (I once read a column by another computer writer, titled "Baud is a four-letter word." This should give you some idea of the level of frustration even knowledgeable PC users feel when confronted with the great baud vs. bps controversy.)

Stated simply, a baud has absolutely nothing to do with your computer or modem. Rather, it's related to how today's telephone lines manage data during a modem connection. As I explained earlier, much of this country's vast telephone network still relies on analog transmission methods. I also discussed the fact that a computer is a digital device—that is, it uses the discrete on/off settings of its electronic switches to represent information. Given this disparity, you might wonder how an analog system, such as a telephone network, can transmit data from a digital device like a PC. The solution to this problem lies with the much-maligned baud.

A baud, named in honor of the French communications pioneer, J.M.E. Baudot, represents the smallest change possible in a signal capable of conveying information. This signal can take any one of several forms: frequency, voltage level, and phase angle, to name three that are commonly utilized in an analog communication channel such as a phone line. A modem's primary function is to convert the digital signals coming from your PC into one of these analog signaling methods.

During a given on-line session, for example, your modem might translate any zeros, or off states, it receives into one frequency. Conversely, when your modem encounters on states, it converts them into a second, different frequency. Once this conversion occurs, a modem is able to transmit the different frequencies through an analog phone line as individual bauds. (The modem on the other end of the connection performs this same procedure in reverse, converting the analog frequencies it receives back into digital signals compatible with the computer to which it is attached.)

In the previous example, each separate frequency burst sent merrily on its way across the phone line corresponds to a single bit of digital data. A modem using this technique to generate 2400 frequency changes per second, therefore, would also be transmitting data at 2400 bps.

In the situation outlined above, a very straightforward, one-to-one relationship exists between baud and bps. Unfortunately, this isn't always the case, for one very critical reason.

You see, there's a practical limit on the number of signal changes a traditional, analog telephone line can handle reliably within a given time period. For a variety of technical reasons, this limit, referred to as the *maximum baud rate*, is 2400 signal modifications per second, or 2400 baud. As anyone who's investigated the modem market knows, however, many manufacturers sell models that seemingly exceed this 2400-baud speed limit. 9600-baud modems are quickly becoming *de rigueur* for anyone truly serious about telecommunications. And modems capable of supporting even higher transmission rates are already available—although, admittedly, the current cost of such cutting-edge technology places it beyond the reach of the average PC user.

But isn't this impossible, you might be thinking, a feat not unlike packing ten pounds of potatoes in a five-pound bag? All I can tell you is to never underestimate the resourcefulness of the computer industry.

The key to shattering the 2400-baud barrier involves finding some method of embedding multiple bits in a single baud. Theoretically, there are many ways to accomplish this. Let me briefly outline one of these, one that's based on a very rudimentary mathematical principle.

In the binary numbering system, where 0 and 1 are used exclusively to indicate off or on switch settings, only four possible combinations can appear in any two-bit sequence:

```
00 01 10 11
```

Now, suppose that, instead of being limited to two frequencies, a modem was designed to recognize four different frequencies, each of which was used to represent one of the binary sequences listed above. Any modem utilizing such a technique could pack two bits into each signal change (baud) sent across a 2400-baud connection, effectively doubling its data transmission rate to 4800 bps. (And before you beat me to a pun, yes, I realize any modem using this technique could be called a two-bit modem.) With a few minor variations on this basic theme, modem manufacturers can realize even more dramatic increases in the number of bits that can be transmitted each second over a 2400-baud phone line.

And that, in a rather convoluted nutshell, is why baud and bps should not be equated with one another. Although they can share the same value, the two are not synonymous. To confuse matters even further (if this is possible), neither one reflects the actual amount of real information traveling across a phone line during a given on-line session. Determining this requires a discussion of several additional telecommunications-related items that will resurface in later chapters, once we begin setting up and using pcANYWHERE.

Understanding data structure

As I pointed out earlier, your PC requires more than a single data bit to represent a specific letter, number, or other character. In the example cited previously, 01000001, the ASCII code for an uppercase *A*, is eight bits long. In the best of all possible worlds, therefore, a modem using eight-bit ASCII codes and operating at 2400 bps would be able to exchange characters with another computer at the paltry rate of 300 (2400 divided by 8) characters per second, or *cps*. Several additional elements used to organize the individual bit settings generated by your PC into real information can depress this figure even further.

Organizing data

True communication involves much more than just stringing together long sequences of characters. In a written document such as this book, for example, spaces are used to divide groups of characters into recognizable words.

Considerhowdifficultitistoreadthisline, for instance, and you'll quickly recognize how much communication depends on some type of organization or structure to be effective. For example, a few well-placed spaces can transform the previous gobbledygook into the following, comprehensible sentence:

Consider how difficult it is to read this line.

pcANYWHERE, indeed, any telecommunications program, employs similar techniques, called *communication parameters*, to organize individual data bits into useful information. The most critical thing to remember about communication parameters is that they must be identical for both computers involved in an on-line session before a successful exchange of data can occur.

You'll learn the precise procedures used to set pcANYWHERE's communication parameters in chapter 4. For now, I simply want to introduce you to several terms relating to communication parameters. Understanding what each represents is critical to being able to use pcANYWHERE efficiently.

Data bits

Data bits define how individual bit settings are organized into discrete characters during a telecommunications session. The *data-bit* setting specifies how many individual bits will be used to indicate a single character. For example, in the previous section, 01000001, the ASCII code for an uppercase *A*, provides an example of a character based on a data-bit setting of 8. The most common data-bit settings used during PC-to-PC telecommunications are 7 or 8.

Stop bits

Stop bits serve a function similar to the spaces between individual words on the printed page, in that they indicate the end of a character comprised of multiple data bits. Stop bits tell the computer on the receiving end of a modem connection, in a manner of speaking, "This is the end of the current character bit sequence. Go ahead and either record or display this character." The most common stop-bit settings for PC-to-PC communications are 1 or 2.

Parity bit

As reliable as today's phone systems might be, they aren't perfect. Originally designed to handle the continuous sound waves that make up the human voice, they require all kinds of electronic juggling to transmit the discrete on/off signals generated by a digital device like your PC. This, you might recall, is the primary reason you must use a modem to begin with. A drop in the quality of a telephone connection—hisses, echoes, and static bursts you hear during phone connections, or what we humans commonly refer to as *line noise*—can wreak havoc on your computer's attempts to converse successfully with another computer.

One way to avoid the types of errors that can result from line noise is to have your communications software attach a parity bit to each character it sends. The software on the other end of the connection can then use this parity bit to verify that the bit pattern of the character it received matches what was transmitted. Today's modems generally rely on one of three parity settings: odd, even, and none.

With error checking enabled, pcANYWHERE uses an extra bit to make the sum of the individual bits in a character correspond to the specified parity. Suppose, for example, that you attempt to transmit a capital *A* in the eight-bit ASCII format mentioned earlier using odd parity. As you've seen several times already, the eight-bit ASCII code for *A* is 01000001. The sum of its individual bits, therefore, would be 2, an even number.

To achieve odd parity, pcANYWHERE would append an extra 1 to the end of this bit pattern and actually transmit 010000011. The software at the other end, which would also need to be set to odd parity, would check to see whether the bit pattern it actually received added up to an even number. If it does, that character would be passed on for additional processing (minus, of course, the parity bit). If the parity check was not successful—that is, the sum of the bit pattern arriving at the receiving end of this transmission turned out to be an odd number—the other computer would assume that an error had occurred and request that this character be retransmitted.

Conversely, if even parity had been specified, a 0 parity bit would have been added to the sample *A*, so that the sum of its individual bits remained an even number. Selecting the third parity option, none, effectively disables data verification at the character level.

Putting it all together

Now that we've discussed each of these items, let's see how they can be combined into a single set of communication parameters used to initiate a successful modem connection.

Suppose, for example, that you wanted to use pcANYWHERE to call another computer. You would specify the telephone number of the modem and follow it with the notation 2400,E,8,1. As you might already have guessed, this sequence indicates that the system in question supports communication parameters of 2400 bps, even parity, 8 data bits, and 1 stop bit. Prior to calling that system, therefore, you would need to set pcANYWHERE to use these same settings. (As I mentioned earlier, we'll get into the precise steps required to accomplish this in chapter 4, "Installing pcANYWHERE.")

If you ever connect to a remote system and see an odd assortment of characters and symbols that makes no sense—a strange phenomenon we high-tech computer types refer to as *garbage*—the most likely culprit is a communication parameter mismatch. Don't panic. (There's that admonition again.) Modifying pcANYWHERE's communication settings to the appropriate values should solve the problem.

The numbers game, revisited

Having outlined the various elements used to organize data-bit sequences into individual characters, let's go back and reexamine how these new items affect data throughput—that is, the actual speed at which real information flows back and forth between the two computers involved in a pcANYWHERE session.

Suppose, for example, that you've configured pcANYWHERE to operate at 2400,E,8,1, the communication parameters mentioned in the previous section. With these settings it would take 10 separate bits to represent each character:

Data bit length	8
Stop bit	1
Parity bit	1
Total bits	10

At 2400 bps, the specified transmission rate, data would move through this modem connection at a maximum speed of 240 cps, or 2400 bps divided by 10 bits per character. (I say "maximum" because additional considerations such as line noise, which might force corrupted blocks of data to be retransmitted based on the results of an erroneous parity check, can reduce this figure to an even lower number.) As the above calculation demonstrates, the bps setting you specify with pcANYWHERE provides only a general indication of how much real information travels back and forth across the phone lines.

Synchronous vs. asynchronous communications

Before moving on to the "nitty gritty" of setting up your system to run pcANYWHERE, let's briefly examine two additional terms you'll encounter in almost any discussion relating to telecommunications, *synchronous* and *asynchronous*.

These terms are used to describe the two different ways that computers can coordinate the exchange of data over a modem connection. Let's discuss synchronous communications first, after which you'll be able to virtually ignore it for the remainder of this book.

Synchronous communications

If nothing else, computers are efficient. With the proper program, a computer can perform a complicated sequence of tasks much more reliably, and with a greater degree of accuracy, than the person running it.

Note: I don't make this statement to denigrate human intelligence or shatter human confidence. Nor should it. Just remember that, for all their positive attributes, computers are also extremely limited; after all, as I explained in our previous discussion concerning binary digits, a computer can count to only 1.

Left to their own devices, two computers can coordinate with great precision the process of transmitting information over a modem link. Suppose, for example, that two computers agreed to synchronize their on-line activities by exchanging one character with one another, and one character only, each second. (In truth, computers work much more quickly than this, operating at speeds measured in something called *nanoseconds*, or one-billionths of a second. Such speeds are almost beyond human comprehension, however. Consequently, and with your indulgence, I'll stick with a one-second time interval in the following discussion.) Under such an arrangement, there's no need to rely on stop bits to identify the individual characters being transmitted. Rather, each computer would know that the information it received during any one-second interval would comprise a single character. The technical term used to describe this synchronized approach to data-transfer is *synchronous communications*.

An on-line session utilizing synchronous communications can be extremely efficient and quite reliable, provided it's managed exclusively by the two computers involved. Add a person to the mix, however, and you need to find some other way to coordinate data transfer, one more forgiving of the all-too-human shortcoming of not being able to function reliably in one-nanosecond time increments.

Asynchronous communications

During asynchronous communications, the two computers involved don't rely on predetermined time intervals to coordinate their on-line activities. Rather, they utilize the data structure outlined earlier (data bits, stop bits, parity bits, etc.) to identify individual characters. Almost all PC-to-PC communication programs, including pcANYWHERE, use the asynchronous method.

Summary

That about does it for the essential groundwork. In the previous chapter, you discovered what pcANYWHERE is, and what it allows you to do. This time around, I described some fundamental concepts of telecommunications, an awareness of which will help you understand how pcANYWHERE allows two computers to communicate with one another. So, where do you go from here? I would think you'd want to prepare your system to run pcANYWHERE. Well, that's exactly what you're going to do, starting in the very next chapter.

3
CHAPTER

Getting ready

Only one more piece of unfinished business to take care of, and then you'll be ready to begin the actual process of installing pcANYWHERE on your system. In this chapter, I'll explain how to set up an internal or external modem to work properly with your PC.

Note: If you're already using a modem with other programs, your initial reaction might be to skip this chapter and jump right into the good stuff. Obviously, I can't glue your eyes to the page. Still, I would recommend that you kick back, kick off your shoes, and kick into "learn mode" for however long it takes to read the following material. In addition to outlining the specific steps required to connect this critical tool of telecommunications to your PC, this chapter also contains general information about modem operations. For example, have you ever wondered what all those flashing lights on the front of an external modem indicate? Read on, and you'll find out; skip ahead to the next chapter, however, and the answer to this burning question will remain forever a mystery.

Installing your modem

As I explained in chapter 1, all modems fall into one of two basic categories, internal and external modems. Connect an internal modem and an external modem to the same computer, and you'll hardly notice the difference between the two. They function almost identically to one another. Installing each, however, entails vastly different procedures. I'll begin by outlining the steps required to install an internal modem on your PC.

Installing an internal modem

An internal modem, as its name implies, is placed inside a computer. Given their "black box" reputation, the idea of removing a PC's case and actually installing a new piece of equipment within it might intimidate you. Don't let it. The people responsible for the original IBM PC—and, by extension, the manufacturers who produce the hundreds of IBM-compatible machines, or *clones*—have designed their products in such a way that adding a new piece of equipment, called a *peripheral*, is a relatively easy undertaking. (In truth, this was one of the primary reasons that the first IBM PC was such an overwhelming success. The original PC/XT might not have been state of the art when it was initially released, but in a moment of rare foresight, the Big Blue engineers built into their new offering the potential to become so, over time.)

PC expansion slots

Almost all IBM-compatible personal computers include an *expansion bus*. Figure 3-1 shows the insides of a fairly typical PC motherboard. The expansion bus in this figure consists of a series of slots running along the back of the motherboard, a special board on which are mounted a system's primary components: its memory chips, BIOS chips, configuration switches, and the like. You can add new features to your PC, or enhance its current capabilities by inserting various *expansion cards* into these slots. The different types of expansion cards that can easily be installed in a PC's motherboard include:

- additional memory, or RAM
- a video display card
- serial ports
- parallel ports
- a bus mouse
- external storage devices (CD-ROM, external hard disk, etc.)
- input devices (scanner, voice recognition card, etc.)
- sound cards
- network cards
- an internal modem card

Although you're going to be interested primarily in the last of these, a modem card, the procedures that follow can be used to install virtually any kind of expansion card in your PC.

Removing the system cover

In order to add an internal modem to a PC, you first must remove its system cover. On most IBM-compatible PCs, this involves removing several screws that are generally located on the back of the main system unit, as illustrated in Fig. 3-2.

Note: Minor design variations exist among different PC systems. Figure 3-2 indicates the location of the cover screws on a typical IBM or compatible desktop unit. Some newer designs replace these standard screws with so-called *thumb-*

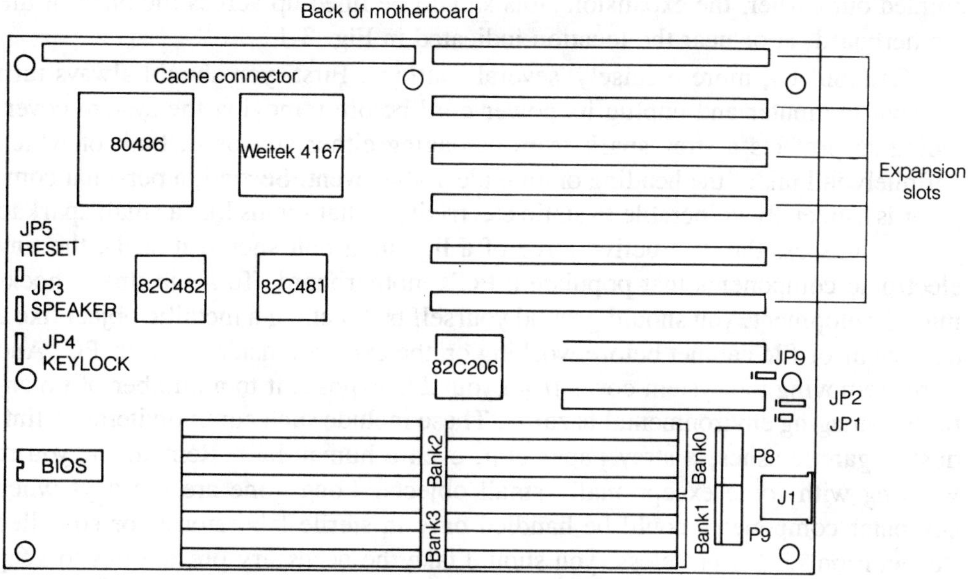

3-1 You install an internal modem using the expansion slots located on your PC's motherboard.

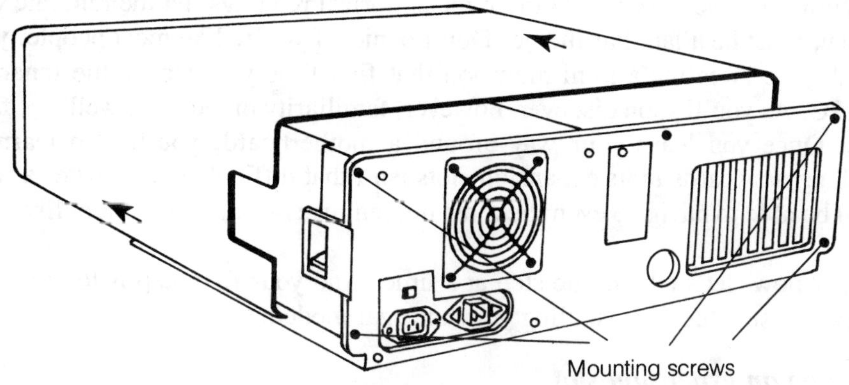

3-2 You must remove the system cover to install an internal modem in your PC.

screws, which can be removed without a screwdriver. Accessing the expansion slots on some tower systems, a relatively recent variation on the original IBM design, often involves sliding off a side panel. The specific procedures used to expose the motherboard of your PC should be outlined in its system documentation.

After these screws are removed, you can remove the system cover by slowly pulling it toward you. (I say *slowly* because you'll want to be careful not to entangle any internal wires or connections on the system cover as you remove it.) Once the system cover has been removed, you'll be able to work inside your PC. As I pointed out earlier, the expansion slots should be lined up across the back of the motherboard, at or near the location indicated in Fig. 3-1.

Caution! Or, more precisely, several cautions: First, you should always turn off your computer and unplug its power cord before removing the system cover. Doing so protects a stray spark from damaging either you or it, both of which definitely fall under the heading of an undesirable event. Second, a personal computer is especially vulnerable to static electricity. What seems like a small spark to you can possess the destructive force of a lightning bolt should it strike the tiny electronic components that populate a PC's motherboard. To avoid this "shocking" development, you should ground yourself by touching a metallic object like a desk chair or file cabinet before working on the exposed insides of your PC. And third, removing the system cover from your PC exposes it to a number of potentially damaging environmental hazards. These include such common items as lint, dust, cigarette ashes, a stray paper clip, even a human hair. Remember, you're working with some exceptionally small objects. Long gone are the days when computer components could be handled only in sterile laboratories or so-called "clean rooms." Nevertheless, you should take the necessary precautions to keep such contaminants away from the delicate components inside your PC whenever its system cover is removed.

Now that I've raised all of these metaphorical red flags, let me reiterate what, by now, must be a familiar theme: Don't panic. If you're like most people, you'll probably be sweating a mini-monsoon that first time you expose the innards of your PC. As you'll soon discover, however, familiarity breeds . . . well . . . familiarity. Once you learn your way around a motherboard, you'll also learn that installing new cards in its expansion slots isn't that difficult a task. After a while, you'll be able to navigate even the most hidden recesses of your system like an old pro.

For now, let's take it one step at a time. And your first step is to identify an empty slot suitable for installing your internal modem.

Selecting an expansion slot

Figure 3-3 shows a typical internal modem card. Notice the row of connectors protruding from the bottom of this card. Take special care not to touch these connectors, which are gold-plated and slide into the expansion slot, while installing your internal modem. They are used to transfer information, in the form of electrical signals, between the modem card and your PC's motherboard. Dust, dirt, or fingerprints can interfere with the electrical signals passing between these connectors and the contacts in the expansion slot.

Different PCs contain different types of expansion slots. Due to the small number of data paths available in the central processing unit (CPU) of the original IBM PC and compatible systems, the expansion slots built into these early models could transfer information to and from the motherboard only eight bits at a time.

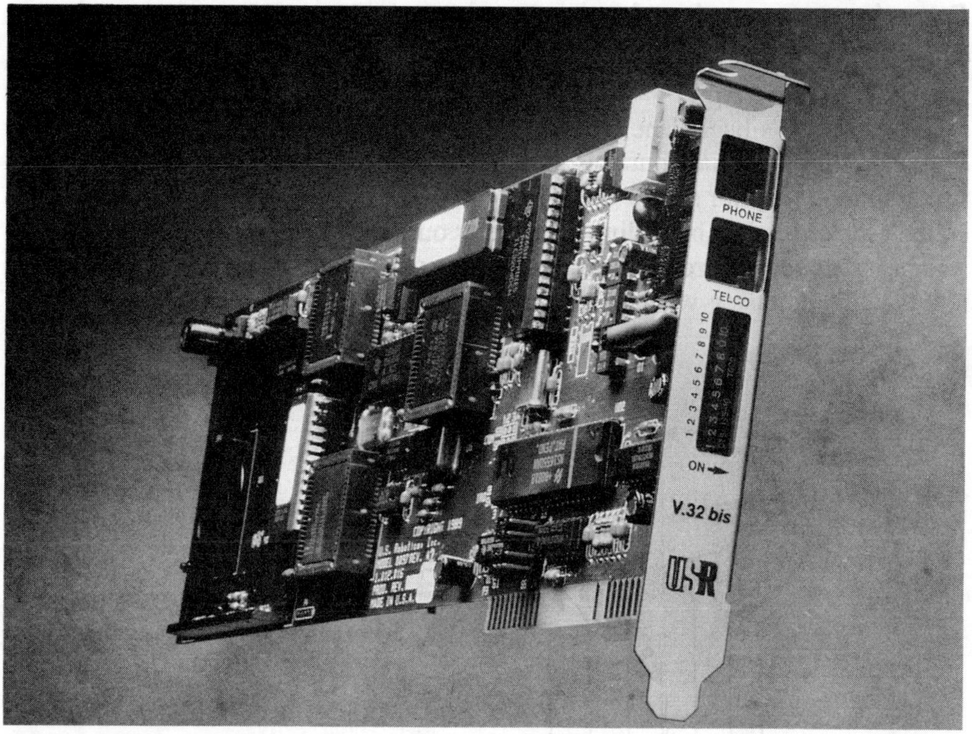

3-3 An internal modem card. USRobotics

This situation improved somewhat with IBM's second-generation (or so-called AT) line of personal computers. Their CPU, an Intel 80286, allowed data to be transferred over an expansion bus 16 bits at a time, or approximately twice as fast as was possible with the earlier PC/XT models. To accommodate this improvement in data-transfer capabilities, the motherboards on these newer systems contain larger expansion slots.

Expansion cards supporting eight-bit data transfer—that is, expansion cards with fewer connectors—will fit in a 16-bit AT slot. The opposite, however, is not true; you cannot insert an AT card in an 8-bit PC/XT expansion slot. (Imagine trying to screw a 1/4″ nut onto a 1/2″ bolt, and you'll have a pretty good idea of why this is so.)

As a rule, a dedicated internal modem card will use the earlier, eight-bit design. Recently, however, several manufacturers have started to sell hybrid models—for example, an expansion card that contains both an internal modem and supports FAX transmissions—as a single card that often requires a 16-bit slot to work properly. Whenever possible, try to install your modem card in the smallest expansion slot with which it is compatible. Doing so leaves the larger, 16-bit slots available for other, more powerful expansion cards that might require these additional connections.

Preparing an expansion slot

Once you've decided into which expansion slot you'll install your internal modem, the next step is to remove its protective plate. These plates, generally secured to a PC's backplane by a single screw, protect unoccupied slots against dust and other potential contaminants. To take off this protective plate, you must first remove its anchor screw (see Fig. 3-4). After removing the anchor screw, lift up the protective plate by sliding it carefully out of the backplane slot into which it was inserted. Make sure to set aside this screw in a safe place. You'll need it in a few minutes, to fasten your modem card onto the PC's backplane.

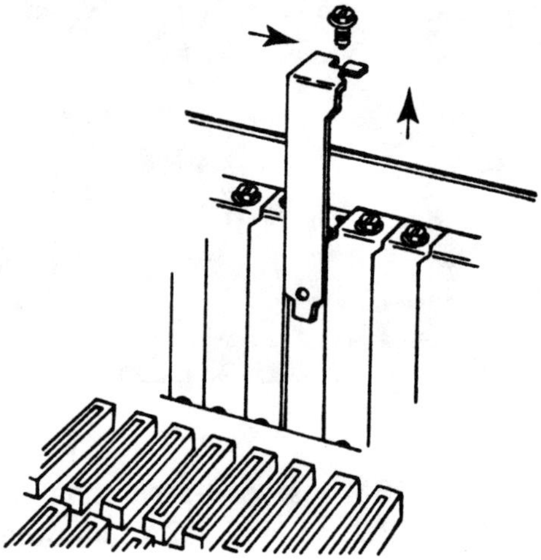

3-4 Removing the protective plate.

Inserting the modem card

With the cover and protective plate removed, you're ready to insert the modem card into the selected expansion slot. The modem card should be mounted on a plate similar to the one you just removed. To avoid damaging the modem card, hold it only by its edges. First, position the modem card's mounting plate against your PC's backplane. Then, carefully slide this plate down into the motherboard. Doing so lines up the the modem card's connector pins with the front section of expansion slot. Finally, press gently against the middle of the modem card until all connector pins are seated firmly in the expansion slot.

Once the modem card is inserted properly, replace and tighten the anchor screw you removed from the original protective plate. Doing so prevents the modem card from slipping out of the expansion bus. This can happen over time if the modem plate isn't firmly anchored to your system's backplane.

Now that everything is in place, you can go ahead and replace the system cover. (As you did when removing the cover, be careful not to entangle it in any wires or connector cables inside your PC.) Once you've reassembled your system, you're ready to plug a standard telephone cord into your new internal modem. I'll explain how to do this later in the chapter. Right now, however, I still have to outline the steps required to connect an external modem to a PC, for those of you who have chosen this alternative.

Installing an external modem

Installing an external modem requires some additional equipment that isn't needed with its internal counterpart. For starters, your system will need to have at least an open external serial port available.

A serial port is a special input/output device your computer uses to communicate with the outside world, so to speak—that is, those peripherals not inserted directly into one of its expansion slots. As a rule, this serial port is itself mounted on a special card that, in turn, needs to be inserted into your system's expansion bus. With some newer PCs, all the electronic circuitry associated with a serial port is built into their motherboard. On such systems, you'll usually find an external connector for this I/O port located on the back of the system cabinet. It might be identified as a serial port, or it could be called COM1 or COM2.

Choosing the right serial cable

You connect an external modem to your system using what is called, appropriately enough, a *serial* cable. The specific kind of serial cable required for your modem depends on what type of I/O port your PC has.

The serial ports on an IBM-compatible PC adhere to one of two designs, as illustrated in Fig. 3-5: 9-pin or 25-pin adapters. If you've already figured out that the only difference between these two connector types is how many pins each contains, give yourself a gold star. Most external modems include a 25-pin connector.

3-5 External modems use either a 9-pin or 25-pin cable connector.

Consequently, you'll require one of the following two cable types to connect an external modem to your PC:

- If the serial port on your PC uses a 9-pin connector, you'll need a serial cable with a 9-pin adapter for the end that connects to your PC and a 25-pin adapter, which can be plugged into the external modem.
- If the serial port on your PC uses a 25-pin connector, you'll need a serial cable with a 25-pin adapter on each end.

Sex and the single cable

Our next topic of discussion is . . . SEX. (That got your attention, didn't it?) PC connectors and cables are gender-specific—that is, they come in both male and female varieties. A connector or cable containing the actual pins used to transmit electronic signals between your PC and an external device is called the *male adapter*. Conversely, if a connector or cable contains the holes into which these pins are inserted, it is referred to as a *female adapter*.

The rule here is simple. If the serial port on your PC has a male connector, as most do, then you will need a female adapter on the end of the serial cable that connects to it. Because most modems are equipped with 25-pin female connectors, you'll need a male adapter on the modem end of the serial cable used to attach an external modem to your PC.

Summing up our short gender-specific conversation, most of you will need one of the following two kinds of cables to connect an external modem to your PC: a 9-pin female to 25-pin male serial cable or a 25-pin female to 25-pin male serial cable.

As is true of every rule, this one will have exceptions. If you understand the difference between a 9-pin and a 25-pin connector, however, and realize that there are male and female permutations of each, you should have no trouble purchasing the appropriate cable for your PC and external modem.

Pass the power, please

Unlike an internal modem, which pulls all the electricity it requires directly from a PC's power supply, you'll need to plug an external modem into an electrical outlet. Modems require low-voltage, direct-current power, as opposed to the 110-volt alternating current generated by a standard outlet, which is why they come with an AC/DC adapter. Setting up an external modem to work with your PC, therefore, is a two-step process:

1. Use the appropriate serial cable to connect the external modem to either your COM1 or COM2 serial port.
2. Use the modem's AC/DC adapter to provide it with electrical current.

Modem lite

Before returning to topics universal to all modems, let's look at one more characteristic unique to external models. An external modem can tell you an awful lot about what's going on during an on-line session if you happen to know where to look. And where is that? At its status lights, which are generally located on an external modem's front panel (see Fig. 3-6).

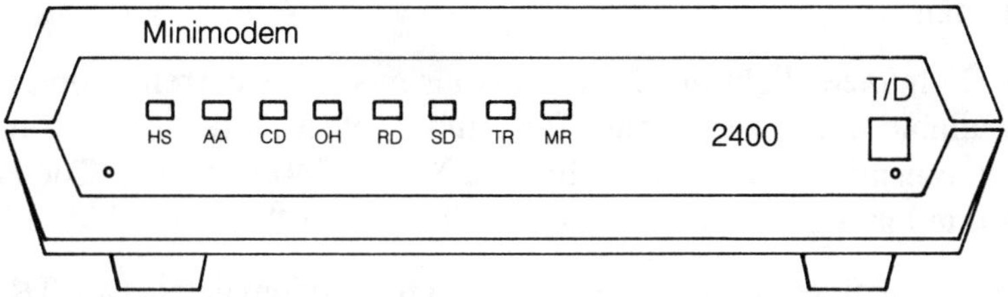

3-6 An external modem's status lights reveal what's going on during an on-line session.

Each time you go on-line, these lights reveal the current conditions for a number of communication parameters or modem operations. Such information can be extremely helpful if you ever have trouble placing calls, installing a modem, or making initial contact with a specific on-line service.

The exact location of a modem's status lights and the order in which they appear will vary from modem to modem. However, each light is usually labeled with a two-character abbreviation, indicating its function. Conditions revealed by a modem's status lights include:

MR The Modem Ready light lets you know that your modem is turned on and ready to operate. If this indicator does not light up when you turn on your modem, check the connections for your power adapter. Make sure it's inserted securely into both the wall outlet and the back of your modem. If the problem persists and you're certain the wall outlet works, you might need to replace your power adapter or, possibly, it might be time to buy a replacement modem.

RD The Receive Data light flickers each time the modem transfers data to your computer. This happens whenever your PC is receiving information of some kind from the remote system.

SD The Send Data light flickers each time your computer transfers data to the modem. This happens whenever you're transmitting information from your PC to the remote system.

AA The Auto Answer light goes on whenever your modem is set to auto-answer mode. With auto-answer in effect, your modem will automatically answer any incoming calls.

CD The Carrier Detect light goes on whenever your modem successfully makes a connection with—detects a carrier signal from—a remote computer. It should remain on for your entire on-line session. At the end of each call, one or the other computers will disconnect its line, at which time the carrier signal is dropped and the CD light goes out.

OH The Off Hook light goes on whenever your modem takes control of the phone line. This is equivalent to taking your telephone receiver off the hook. Hence, its name.

HS The High Speed light indicates that your modem is currently operating at its highest available transmission rate. Remember, most modems can communicate at multiple transmission rates—300 bps, 1200 bps, 2400 bps, etc. The HS indicator lets you tell at a glance if your modem is running "full speed ahead."

TR The Terminal Ready light goes on when the modem detects a DTR (data terminal ready) signal from your communications software. This signal informs your modem that a communications program is loaded and ready to run.

Knowing the function of each of these lights can be useful, should you experience problems using pcANYWHERE. For example, if the RD (Receive Data) light does not flicker during a file transfer, that would be a good indication that, for some reason, the other computer system has stopped sending data, and the transfer is either over or else it has been interrupted. If, on the other hand, you're sending a file to another computer and the SD (Send Data) light stops flickering, that means your PC has stopped transmitting data to the other system.

Connecting your modem to a phone line

If you've performed the preceding steps properly, your modem, be it an internal or external model, will be almost ready to use. All that remains now is to actually hook it up to a telephone line.

I'm going to assume here that the wall plugs in your home or office use a modular phone jack (also called an RJ11 plug), like the one shown in Fig. 3-7. If this isn't so—as might be the case for some third-party office telephone systems—you'll probably need a special conversion cable to tie your modem successfully into the phone line. Check with the company that provides your service to see if they offer this equipment. If you don't need a special converter, simply insert one end of an RJ11 cord into the modem, then connect the other end to your modular wall jack. Easy, isn't it?

Note: If your modem includes two RJ11 plugs, one marked *line* and a second marked *phone*, you can "daisy chain" both the modem and a standard telephone (or answering machine, or FAX machine, or almost any other phone-based communications equipment) off a single modular wall jack. To do so, simply use one RJ11 cord to connect your modem to the wall jack, inserting it into the plug

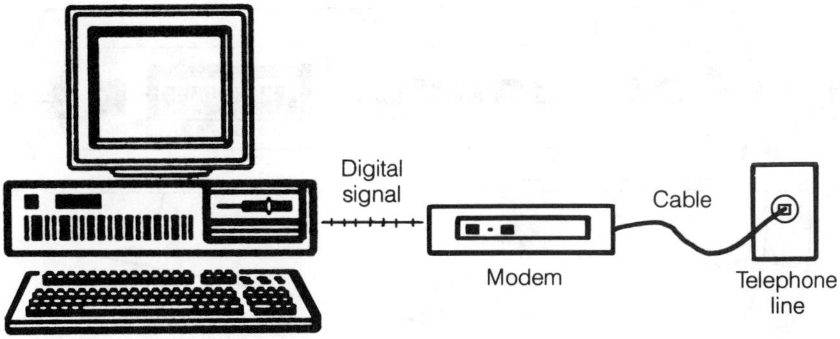

3-7 Your modem plugs into the phone system using a standard RJ11 cord.

marked LINE. Then plug your regular telephone (or other piece of equipment) into the second RJ11 connector on your modem, the one marked PHONE. Daisy chaining equipment in this manner allows your phone to function just as if it were plugged directly into the wall jack.

Configuring your modem

At this point, regardless of whether you own an internal or external modem, it's ready to use. Well, almost ready. It might still be necessary to fine tune your modem's performance to match your PC hardware.

To accomplish this, many modems contain DIP switches—a tiny row of on/off switches that resemble little teeth. Where these switches are located and what they accomplish vary from modem to modem. Figure 3-8 shows a modem that has its DIP switches located under a removable cover on its front panel. An internal modem's DIP switches, on the other hand, might be located either on the modem board itself or mounted so that they can be accessed through the rear mounting plate. The latter arrangement tends to be more convenient because it doesn't require that you remove the modem board to adjust its default configuration.

The setting of a modem's individual DIP switches determines such items as how it responds to commands during an on-line session, the speed at which the modem normally operates, and so forth. How many switches there are and what each one does varies from modem to modem. Within these variations, however, certain consistencies do exist, specifically in the configuration options the DIP switches allow you to specify. For example, virtually all modems will have a DIP switch that allows you to specify whether they should be configured to run in synchronous or asynchronous communication mode. (As I mentioned in the previous chapter, pcANYWHERE runs in asynchronous mode.) A second option generally specified with a DIP switch is whether or not you want to enable, as a default

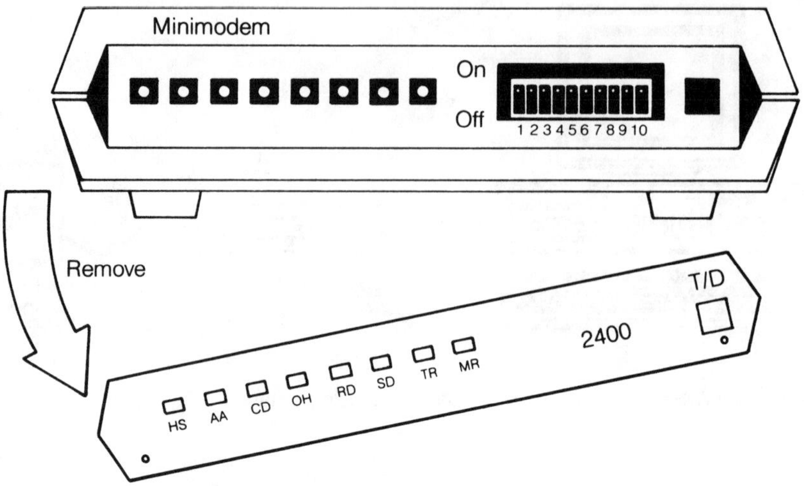

3-8 You use DIP switches to modify your modem's default settings.

setting, *auto-answer*—that is, whether the modem should automatically answer all incoming calls.

Summary

For the past three chapters, I've been providing you with general information relating to telecommunications. The various topics I've covered will come in handy as you install and begin using pcANYWHERE. Now, however, you can break open your pcANYWHERE disk package and power up the ol' PC, because that's exactly what you're going to be doing as soon as you begin the next chapter.

4
CHAPTER

Installing pcANYWHERE

As I pointed out at the end of the previous chapter, it's time to begin applying all the general information you've been receiving up to this point and actually prepare pcANYWHERE to run on your PC. Ultimately, in fact, you're going to install two different pcANYWHERE modules on two separate computers, a host and a remote system. The pcANYWHERE package contains all of the disks and programs you'll need for the following exercises.

Host vs. remote, a brief review

Before getting started, however, let's review briefly the role that each of these computers plays within the pcANYWHERE scheme of things:

- The remote system is the PC you'll use to call up and take control of another computer during a remote computing session. It's the system on which you'll install the pcANYWHERE AWREMOTE.EXE program, using the distribution disk or disks marked *Remote Diskette*.
- The host system is the computer at the other end of this connection. Commands entered at the remote system are actually executed on the host system during a remote-computing session. AWHOST.EXE, the program you'll install on your host system, can be found on the pcANYWHERE distribution disk or disks marked *Host Diskette*.

Note: Although I'll continue to refer to the PCs on which you'll be installing the two pcANYWHERE modules as separate computers, a host and remote system, this need not necessarily be the case. There's no reason you can't install both components of pcANYWHERE on the same PC. (Of course, doing so tends to defeat the whole idea of remote computing, doesn't it?)

Choosing the appropriate distribution disks

When you open your pcANYWHERE package, you'll discover two separate sets of distribution disks: 720K, 3.5" and 360K, 5.25". How many and which of these distribution disks you'll need in order to install the individual pcANYWHERE modules depends on what type of disk drives your computer contains. Both sets of diskettes include the same material—that is, the programs required to install pcANYWHERE on your host and remote systems.

Write protecting your pcANYWHERE distribution disks

Whenever possible, it's a good idea to install a PC program using duplicate copies of its distribution disks. This protects the original disks against accidental damage, thus guaranteeing that they'll be available should you ever need them to reinstall pcANYWHERE at a later date or, alternately, on still another system. The easiest way to create working copies of any disk is with the DOS DISKCOPY command. Before doing so, however, take some precautionary measures to ensure that no harm comes to the original pcANYWHERE distribution disks. Both 5.25" and 3.5" disks provide write protection to prevent important files from being accidentally erased.

Write protecting 5.25-inch disks

Figure 4-1 shows a typical 5.25" floppy disk. Notice the small notch near its top right-hand corner. When this write-protect notch is uncovered, you can record new data on the disk. Covering this notch write protects the disk by preventing DOS from writing any new data to it.

As a rule, boxes of 5.25" disks include several small, adhesive tabs called write-protect tabs. To guard a 5.25" disk against accidental erasure, simply take one of these tabs and cover the write-protect notch, as shown in Fig. 4-2.

Write protecting 3.5-inch disks

Figure 4-3 shows a typical 3.5" disk—a type of disk you'll sometimes see referred to as a *microfloppy*. On 3.5" disks, the write-protect mechanism is built right into the disk's plastic case. Consequently, you don't need a special tab to write protect a 3.5" disk.

Potentially confusing is the fact that write protection on a 3.5" disk works on a principle exactly opposite that of its 5.25" counterpart. A microfloppy is write protected when its write-protect hole is uncovered—that is, when you slide the

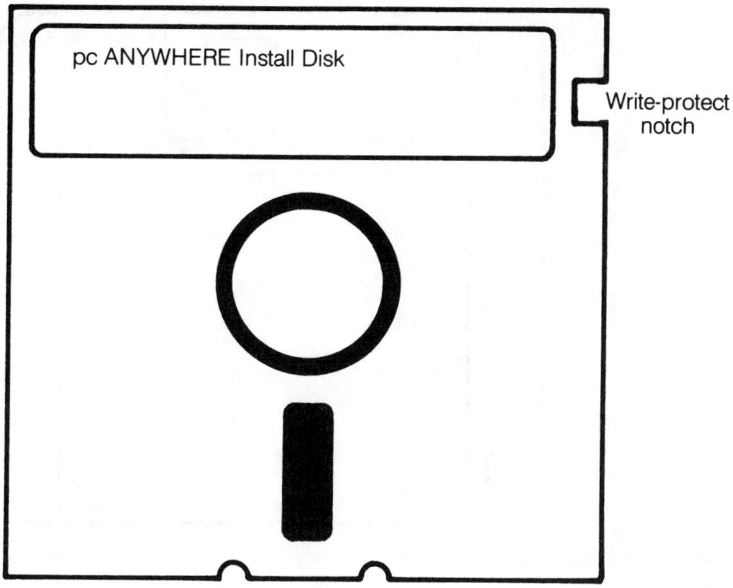

4-1 5.25″ disks include a write-protect notch.

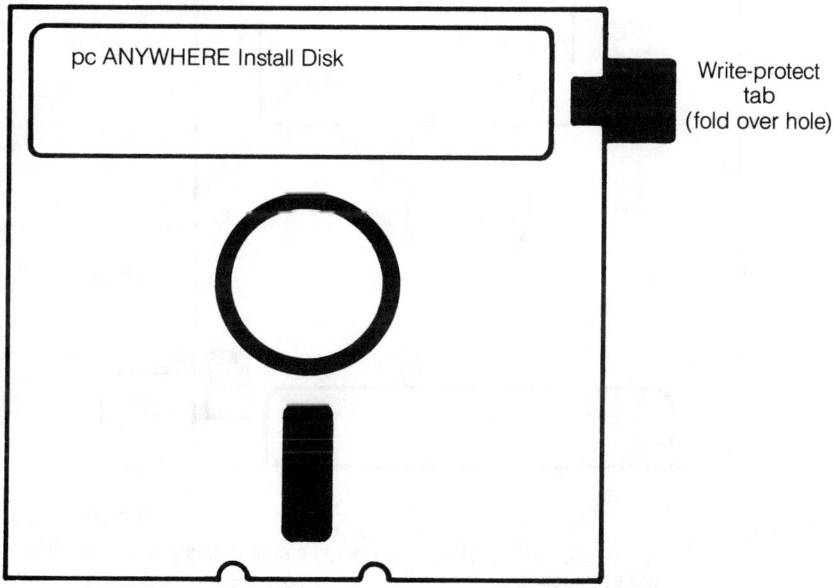

4-2 Covering the write-protect notch with a special tab protects files stored on a 5.25″ disk against accidental erasure.

built-in plastic tab back and expose the write-protect hole, as shown in Fig. 4-4. Covering this hole defeats write protection and allows you to copy new data to the disk.

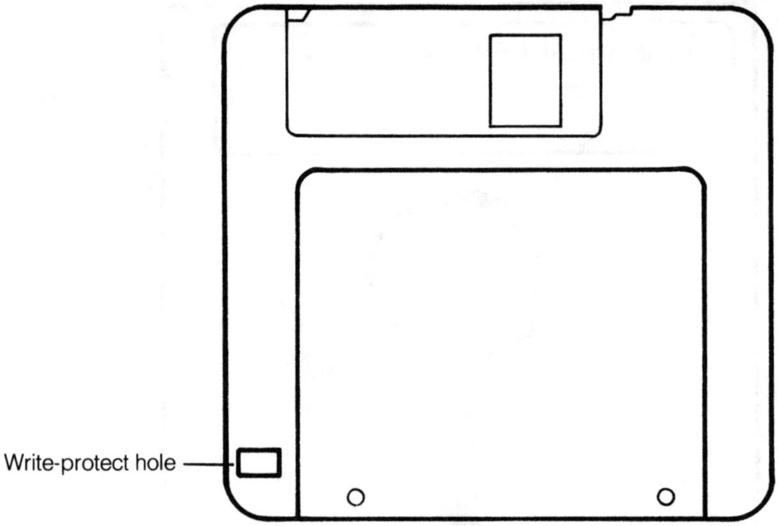

4-3 Write protection is built into a 3.5″ disk.

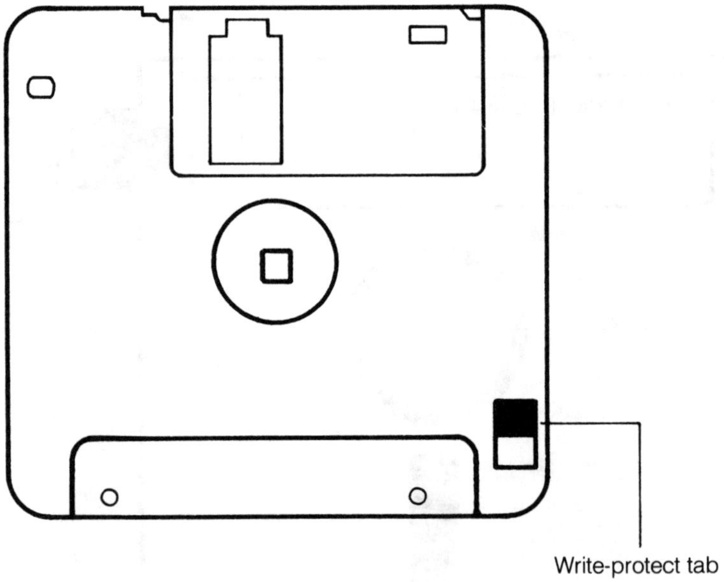

4-4 To record data on a 3.5″ floppy disk, you must slide the plastic tab over its write-protect hole.

Before creating the working copies you'll use to install pcANYWHERE, go ahead and write protect the original distribution disks, as a safeguard their being accidentally erased or damaged. To write protect your pcANYWHERE distribution disks, place a write-protect tab over the notches on the 5.25″ pcANY-

WHERE distribution disks or slide back the plastic write-protect tab built into the pcANYWHERE 3.5″ distribution disks.

With this little bit of preventive maintenance out of the way, you're now ready to begin using the DISKCOPY command to create working copies of your pcANYWHERE distribution disks.

Duplicating a floppy disk

While dependable, no computer is perfect—neither are any computer programs, nor the people who use them, for that matter. Though such things happen only rarely, it's possible to damage a floppy disk. An even more common occurrence is the accidental erasure of files on a disk by an ill-timed or mistakenly entered COPY or FORMAT operation. For this reason, as I stated earlier, you should always make working copies of any distribution disks you own—that is, exact duplicates of the original disks. The best time to do this is before you attempt to install a new program (in this case, pcANYWHERE) on your PC. DOS includes a special program, DISKCOPY, for precisely this purpose.

DISKCOPY vs. COPY

If you've been using your PC for any length of time, chances are you're already familiar with the DOS COPY command, which allows you to copy individual files from one disk to another or, alternately, between different directories on the same disk. DOS's DISKCOPY command works much like a standard COPY operation, except that it produces an exact duplicate of one disk on a second disk of the same type and size. In essence, DISKCOPY creates a "clone" of any disk it duplicates. For this reason, DISKCOPY is the ideal tool for creating working copies of your original pcANYWHERE distribution disks.

Running DISKCOPY

How you enter the DISKCOPY command depends on whether you have a single- or double-drive system—that is, whether your PC includes one or two floppy disk drives, respectively. To run DISKCOPY on a single-drive system, type DISK COPY A: A: and press Enter. To run DISKCOPY on a dual-floppy system, type DISKCOPY A: B: and press Enter.

Caution: The second method shown above will work with only two disk drives of the same size. If your system's A and B drives are designed to hold the two different sizes of disks, you'll need to use the single-drive method—that is, type DISKCOPY A: A: and press Enter, just as you would if your PC only has one disk drive.

After loading the DISKCOPY program into memory, you'll be prompted to insert a source and target diskette in the appropriate disk drive, as shown in Fig.

```
A:\>DISKCOPY A: B:

Insert SOURCE diskette in drive A:

Insert TARGET diskette in drive B:

Press any key when ready . . .
```

4-5 DISKCOPY displays instructions for its use.

4-5. Note: If you're working on a single-drive system, DISKCOPY will begin by telling you to insert the source disk in drive A.

The *source* disk is the disk you're copying from—in this case, one of your original pcANYWHERE distribution disks. Conversely, the *target* disk is the disk you're copying to—usually a blank, unformatted disk. With this in mind, on a single-floppy system place one of your pcANYWHERE distribution diskettes in drive A and type DISKCOPY A: A:. If you have a dual-floppy system, place one of your pcANYWHERE distribution disks in drive A and a blank diskette in drive B, type DISKCOPY A: B:, and press Enter.

This starts the DISKCOPY operation, during which DOS creates an exact duplicate of whichever pcANYWHERE distribution disk you placed in drive A on a second disk in either drive A or B. Note: On a single-disk system, you'll be prompted to remove the source disk from drive A and replace it with your target disk at the appropriate time.

After duplicating your first pcANYWHERE distribution diskette, DISKCOPY will ask if you want to copy another disk. Answering Yes to this prompt will start a second DISKCOPY operation. Repeat this process, using a different pcANYWHERE distribution diskette each time, until you've created a working copy of each of your original pcANYWHERE distribution diskettes.

Tip: Once each pcANYWHERE distribution disk is duplicated, you should remove it from the source drive and store it in a safe place. In the next section, you'll use the working disks DISKCOPY created to install pcANYWHERE. Should one of your working copies ever be damaged or erased, you can always run DISKCOPY again to recreate it—an option that wouldn't be available if you were using your original pcANYWHERE distribution disks.

Installing the pcANYWHERE host software

Now that you've created working copies of the pcANYWHERE distribution disks, you're ready to begin installing software on your PC. Start by setting up a host system that you'll be able to take over and control using the pcANYWHERE remote module.

Selecting a host system

But which system should this be? To be honest, the answer to that question depends on how you ultimately plan to use pcANYWHERE. The following scenarios describe just some of the ways in which pcANYWHERE can be used. For each situation, I've identified the appropriate system on which to install pcANYWHERE's host and remote modules.

- Office personnel who often take work home can use pcANYWHERE to provide immediate access to programs and data files stored on their office PC in the evenings or on weekends. If this scenario describes your situation, you would want to install pcANYWHERE's host module on the desktop PC in your office and the remote software on your home computer.
- Using pcANYWHERE's remote-computing capabilities, people whose work entails travel can carry a lightweight portable computer (a laptop or notebook PC) with them, but still have access to all the power of a desktop PC located in their home or office. In this case, you would install pcANYWHERE's host software on the desktop PC and use the remote module on your laptop or notebook computer.
- To reduce the turnaround time required to fill customer orders, the owner of a small business might decide to set up a system where sales personnel in the field can call their orders into a personal computer at the company's main office. In this case, a single copy of the pcANYWHERE host software would be installed on the PC located at the main office. Each salesperson, however, would need to have a separate, licensed copy of the pcANYWHERE remote software for his computer.
- A systems consultant can use pcANYWHERE to access her clients' computers. Doing so makes it possible to diagnose and, in some cases, correct problems without needing to schedule an inconvenient and potentially expensive on-site visit. For this scenario, only a single copy of the pcANYWHERE remote module would be required, which would be installed on the consultant's own system. However, a separate, licensed copy of the pcANYWHERE host software would need to be installed on each client's PC.

Note: Symantec sells special editions of pcANYWHERE, called pcANYWHERE Host and pcANYWHERE Remote, that contain only the host and remote pcANYWHERE modules, respectively. The final two situations described above would require purchasing the appropriate number of one or the other of these special versions of pcANYWHERE.

Certainly, additional uses for pcANYWHERE exist. In truth, the potential applications of pcANYWHERE's remote-computing capabilities are virtually infinite. For your purposes here, the key is to figure out which of your two systems is the appropriate one on which to install the pcANYWHERE host software.

Stated simply, it's the computer you want to be able to call up and take control of from a remote location. (Remember my metaphor of the host at a party, which I described in chapter 1. The host system in a pcANYWHERE session is the computer that opens an electronic door, so to speak, and invites a person operating the remote system in for a visit.)

Starting the host installation program

The pcANYWHERE Host distribution disks include a special utility, INSTALL .EXE, which you'll use in order to install the pcANYWHERE host software on your PC. The disk you'll need to start this utility depends on the type of disk drive you'll be using to install pcANYWHERE.

If you're installing pcANYWHERE from a 5.25" disk drive, INSTALL.EXE will be on a 5.25" floppy disk marked *Host 1 Diskette*. If you're installing pcANYWHERE from a 3.5" disk drive, INSTALL.EXE will be on the 3.5" microfloppy marked *Host Diskette*.

Caution: It's imperative that you run the INSTALL.EXE utility, even if you plan to install your working copy of the pcANYWHERE host or remote software on a floppy disk—as you might do, for example, if you're going to be using pcANYWHERE on a laptop or notebook computer. As part of the pcANYWHERE installation procedure, INSTALL.EXE expands several files that are compressed on the distribution disks to preserve space. For this reason, you can't create a working copy of pcANYWHERE by simply copying files from the distribution disks to another disk.

To start INSTALL.EXE and begin installing pcANYWHERE on the computer you've selected as your host system:

1. Place the appropriate host diskette in drive A of the host system.
2. Type A: and press Enter to make this the active disk drive.
3. Type INSTALL and press Enter.

After a few seconds, your screen will resemble Fig. 4-6. This opening screen lists the options available to you during installation of pcANYWHERE/Host (the collective name for the various programs and utilities that enable a computer to function as the host system). Before getting into what each of these options represents, however, let's examine one extremely useful feature of pcANYWHERE that's available at almost any time during a remote-computing session.

A helping hand

pcANYWHERE includes an extensive, context-sensitive Help system, which you can access at virtually any time during a remote-computing session. The type of information Help reveals depends on what you're doing at the time—which is what's meant by *context sensitive*. To access pcANYWHERE's on-line Help, press the F1 function key.

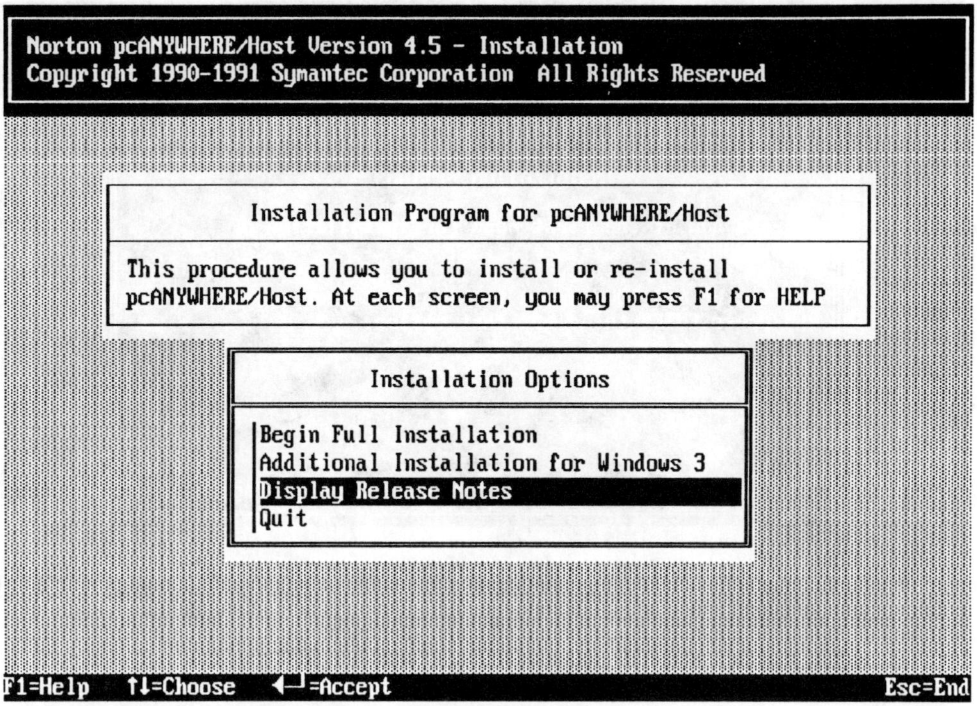

4-6 The opening pcANYWHERE/Host installation options screen.

Because you were just beginning installation, accessing Help at this time will call up general information about selecting installation options, as shown in Fig. 4-7. You'll learn what each of these options accomplishes in a moment. For now, familiarize yourself with the appearance of the pcANYWHERE Help screen.

Sometimes the Help needs more than a single display window to fully explain a specific operation or program feature. Whenever this happens, the following prompt will appear across the bottom of the initial screen:

Page 1 of *x*

where *x* indicates the total number of pages in the current Help message. This is the case, for example, in Fig. 4-7, which indicates that there are three separate display screens associated with the initial installation Help message. To access the next part of the Help message, press PgDn.

This displays the next page of the initial pcANYWHERE installation Help message, as shown in Fig. 4-8. Continued pressing of the PgDn key will call up additional message screens until all Help information associated with the current function has been displayed.

Tip: As you might suspect, the PgUp key performs the opposite function by allowing you to recall previously displayed portions of the current Help message.

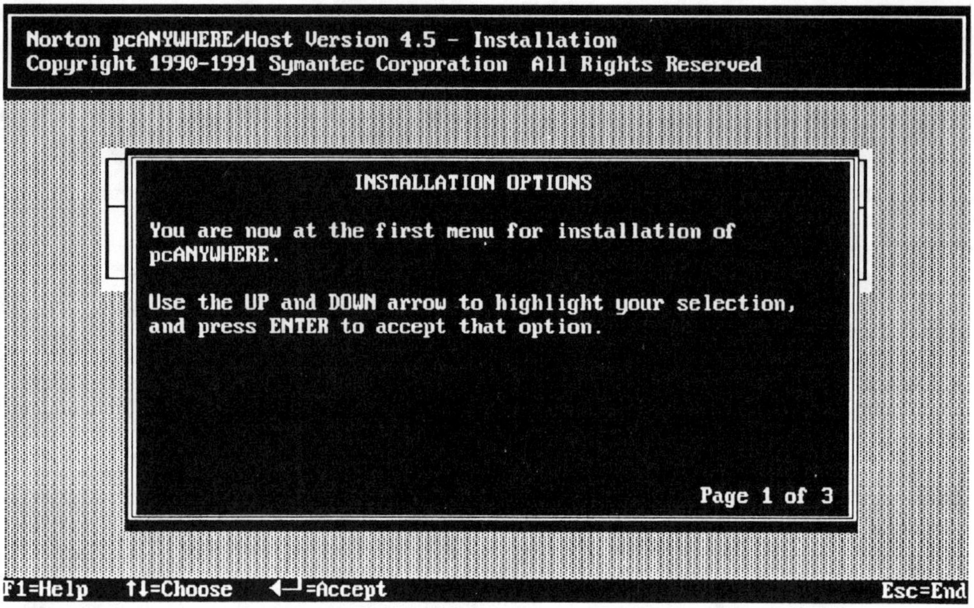

4-7 Pressing F1 displays pcANYWHERE's on-line Help messages.

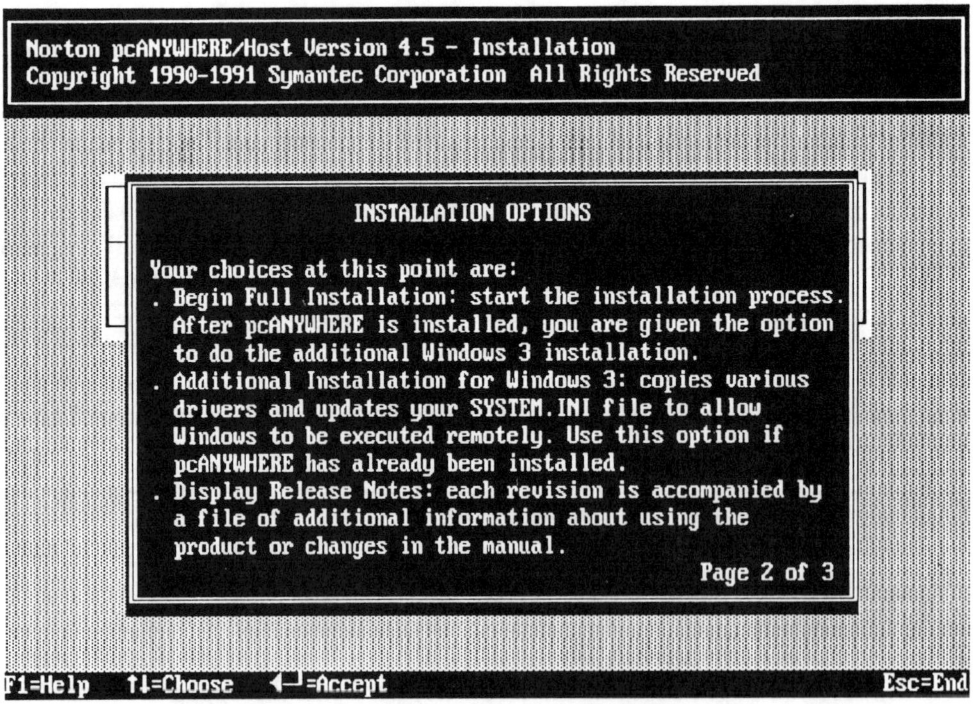

4-8 Use the PgUp and PgDn keys to page through multiscreen Help messages.

The following procedures, therefore, can be used to access Help at virtually any time during a pcANYWHERE session:

- Pressing F1 displays an initial context-sensitive Help message.
- PgDn advances you to the next page of a multi-screen Help message.
- PgUp recalls previous pages of a multi-screen Help message.

For now, let's exit the Help function and continue installing the pcANYWHERE host software. To exit Help, press Esc. As shown in Fig. 4-6, the initial INSTALL .EXE screen offers several options, which include:

- Begin Full Installation
- Additional Installation for Windows 3
- Display Release Notes
- Quit

Because this is the first time you're installing the pcANYWHERE/Host programs, let's specify the Full Installation option. (Windows users, don't worry. The Full Installation option provides you with an opportunity to set up pcANYWHERE to let you run Windows applications on the host system during a remote-computing session.) To select Full Installation, use the arrow keys to highlight Begin Full Installation and press Enter.

Providing installation information

During installation, INSTALL.EXE requests specific information about how you want pcANYWHERE set up to run on your host system. The first thing it needs to know is where you want to store the pcANYWHERE/Host program files. Generally, this will be a dedicated directory path on the hard disk of your host system. To specify a path for your pcANYWHERE/Host files, type C:\AWH and press Enter.

Note: To specify a different drive or directory, simply substitute the appropriate information in the above step. For example, if you want the pcANYWHERE/Host files installed on drive D, you could specify D:\AWH.

Next, INSTALL.EXE requests that you enter a licensee name—that is, the person or company authorized to use this particular copy of pcANYWHERE/Host. To identify a licensee, type in the name of the person or company authorized to use this copy of pcANYWHERE/Host and press Enter.

Note: Like most computer software, each copy of pcANYWHERE is protected by laws designed to help manufacturers fight unauthorized duplication and use of their products. This explains why, in the scenarios presented previously, I refer to the possible need to purchase multiple copies of the pcANYWHERE host or remote modules, depending upon the manner in which the program is being used. Let me make this perfectly clear—software piracy is nothing more than a fancy name for theft. It's illegal and, ultimately, raises the cost of programs purchased by honest, legitimate users. Besides, having to add a "signature" to your copy of pcANYWHERE during installation is a lot less inconvenient than some of the exotic copy-protection schemes employed by software manufacturers in the early days of the PC revolution. So go ahead and cooperate with Symantec on this one. After all, if you're installing a legal copy of pcANYWHERE, what have you got to lose? (I'll step down off my soapbox now.)

Before proceeding, INSTALL.EXE asks you to verify the drive and directory you specified earlier as the target location for your pcANYWHERE/Host program files (see Fig. 4-9). Selecting No at this point will terminate the installation procedure and provide you with an opportunity to return to the DOS system prompt. As far as I can tell, however, you're ready to go. So let's do it.

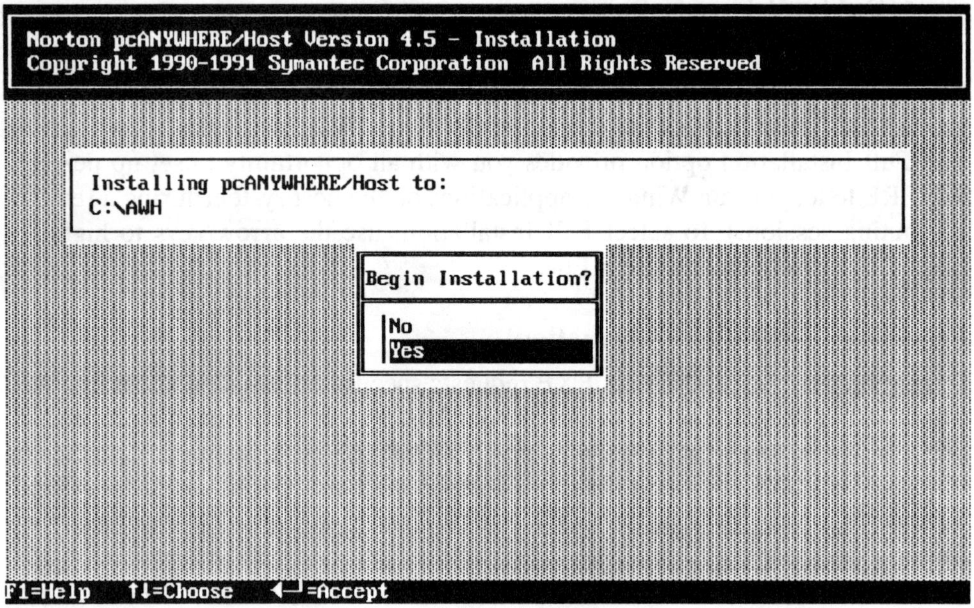

4-9 pcANYWHERE prompts you to verify the specified installation path.

To have INSTALL.EXE actually begin installing the pcANYWHERE/Host program files on your system, press Enter to specify Yes. At this point, INSTALL .EXE will go about its appointed task, requiring little or no intervention from you. (If you're installing pcANYWHERE from 5.25″ disks, you'll be asked to insert the second distribution disk in drive A at some point during the process.) Basically, INSTALL.EXE is expanding files originally compressed on the distribution disks and copying them to the specified disk and directory. Once this process is completed, you'll see the following message:

Installation Complete
Press <Enter> to continue . . .

When the prompt appears, press Enter.

pcANYWHERE does windows

Yeah, yeah, I know. The old "Do you do windows?" line is a stale joke. However, it does provide me with the perfect lead-in to the impressive new feature that was added to pcANYWHERE beginning with version 4.5—The ability to remotely control Windows applications. Answering Yes to the next prompt in the installation sequence (see Fig. 4-10) causes INSTALL.EXE to configure your host system so that someone with the pcANYWHERE remote software can run Windows, Microsoft's popular graphical user interface (GUI) and the current apple of the MS-DOS world's eye.

4-10 You can install pcANYWHERE/Host to support remote access to Windows applications.

Given Windows' graphics-based design, this is no small accomplishment. It's one thing to transmit character-based screens and keyboard input over a modem connection. In most cases, this can be accomplished through the relatively simple technique of intercepting BIOS (basic input/output system) calls and transmitting them across a phone line to the other computer involved in a remote-computing session. Coordinating a graphics-based environment such as Windows over a remote connection, however, is infinitely more challenging.

For one thing, Windows' already lethargic performance would tend to slow down to a virtual standstill unless the software coordinating the remote session takes steps to compensate for the potential bottleneck created by Windows' bit-mapped graphic screens and mouse-driven user interface. pcANYWHERE accomplishes this by editing the Windows' SYSTEM.INI configuration file to substitute its own display, keyboard, and mouse drivers for the ones initially shipped with Windows. As you might suspect, therefore, pcANYWHERE won't be able to configure a host computer for remote Windows operations if it can't find this file—that is, unless Windows is already installed on the host system.

For this reason, you have two options when pcANYWHERE displays the prompt shown in Fig. 4-10:

- If Windows hasn't been installed on your host system, use the arrow keys to highlight No and press Enter.
- If Windows is installed on your host system and you want to be able to run Windows applications remotely, highlight the Yes option and press Enter.

Selecting No will complete installation of the pcANYWHERE host software and returns you to the DOS system prompt without enabling Windows support. (Should you add Windows to your host system at some later time and want to modify pcANYWHERE to support remote Windows operations, simply rerun INSTALL.EXE and select Additional Installation for Windows 3 when you reach the options screen shown in Fig. 4-6.)

Choosing Yes at Fig. 4-10 causes INSTALL.EXE to display the message screen shown in Fig. 4-11, requesting additional information about your Windows

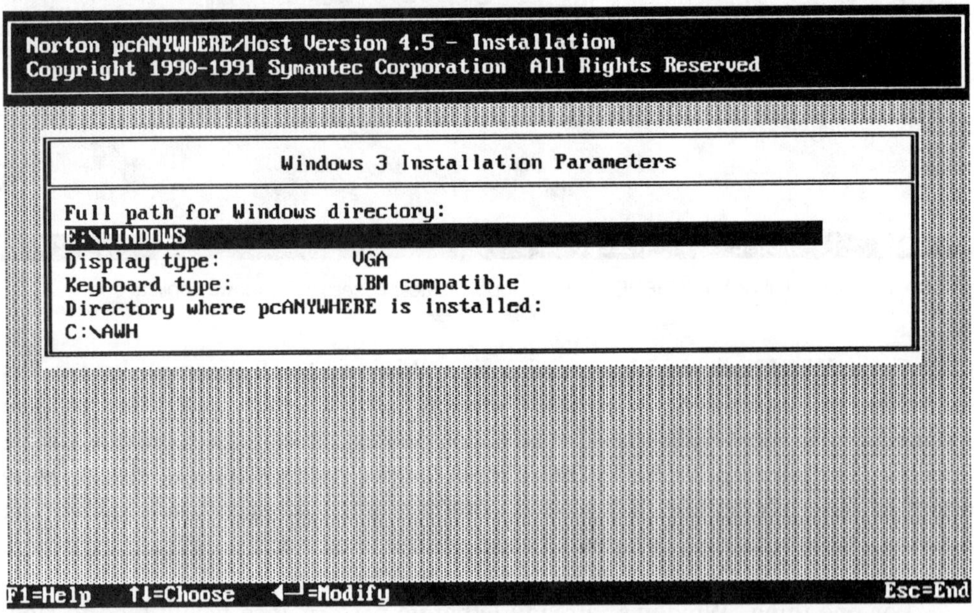

4-11 The installation utility asks you to verify its analysis of your hardware prior to modifying the Windows environment on your pcANYWHERE host system.

setup. If necessary (although it probably won't be; INSTALL.EXE does a very good job of analyzing your PC hardware), you should modify the values listed on this screen to reflect conditions on your host system—that is, the computer on which you're currently installing the pcANYWHERE/Host software. When all the information in Fig. 4-11 is correct, press Esc to advance to the next stage of Windows 3 installation.

INSTALL.EXE will then propose to modify the Windows environment so you'll be able to use pcANYWHERE to control Windows applications over a remote connection, as shown in Fig. 4-12. At this point, you have four options:

Make the changes Selecting this option tells INSTALL.EXE to automatically make any required changes to the Windows SYSTEM.INI file. (Your original SYSTEM.INI file will be renamed SYSTEM.AW0.)

View changes only Selecting this option displays the changes that would have been made, but INSTALL.EXE doesn't actually edit SYSTEM.INI. (You can, if you like, print out or copy down the proposed changes and then make them manually, using any standard text editor.)

Send output to other file Selecting this option creates a new file containing what would have been the edited version of SYSTEM.INI. Windows will use its

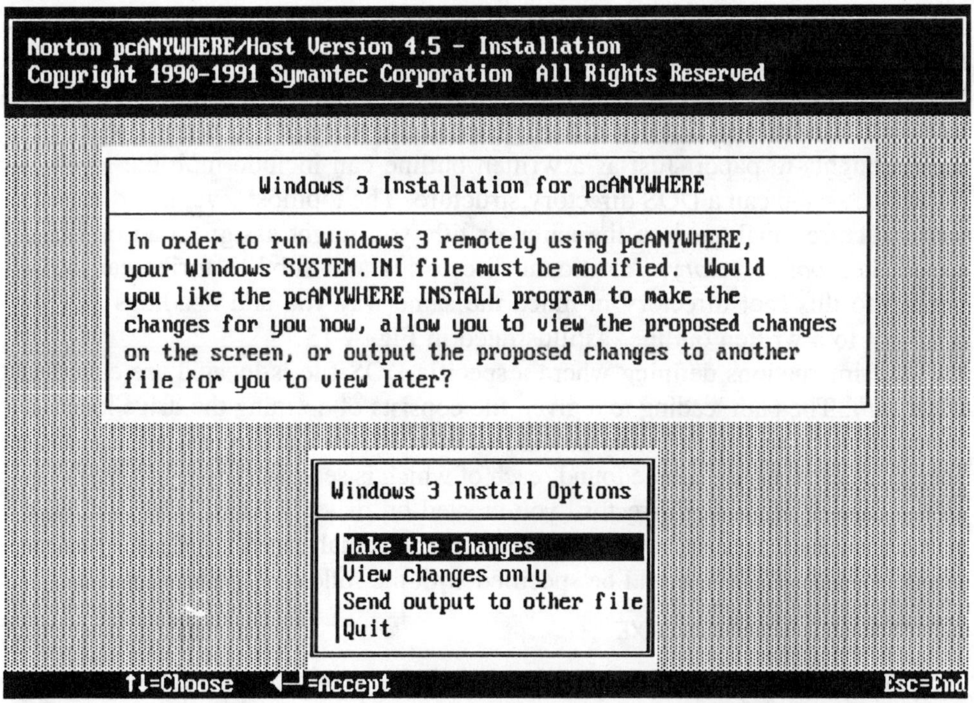

4-12 INSTALL.EXE can automatically modify Windows SYSTEM.INI to support remote access.

original SYSTEM.INI file (and, therefore, be incompatible with pcANYWHERE) unless and until you incorporate the modifications in this new file into your SYSTEM.INI file.

Quit Selecting this option terminates INSTALL.EXE without setting up your host system to allow Windows applications to be run remotely. (The rest of your pcANYWHERE installation remains in place, however.)

To configure pcANYWHERE for Windows compatibility, press Enter to choose the default option and make the changes. To select another option, use the arrow keys to highlight the desired action and press Enter. Note that you should be using at least a 9600-bps modem with Windows.

At this point, installation of pcANYWHERE/Host is almost finished. Before closing down shop, however, INSTALL.EXE will display a message suggesting that you add the directory you created for the pcANYWHERE host programs to your DOS search path. That message refers you to the DOS documentation for an explanation of how your PC uses search paths. But, what the heck; I'm feeling generous. I'll throw in a brief explanation of DOS search paths here to save you from having to wade your way through one more dry and unintelligible software manual.

DOS search paths

Search paths provide a way for DOS to find and execute programs stored on a different disk or within a different directory from the one that's currently active. One way to picture the DOS directory structure is to compare it with a written outline—a common organizational tool people have used for centuries to commit their thoughts to paper. Just as a written outline can include multiple levels, or *headings*, so too can a DOS directory structure. The topmost level in a disk directory structure, analogous to the name or title you might assign to an outline, is called the *root directory*. Additional directories and subdirectories can be appended to this root directory in much the same way you add headings and subheadings to a written outline, as illustrated in Fig. 4-13.

The instructions defining where a specific DOS file is located are called that file's *path*. The path leading to a given file consists of defining the drive letter, followed by a colon, then listing any directories and/or subdirectories required to specify where that file can be found, each of which is separated by a backslash (\ ). In the case of the AWH directory you created on drive C when you installed the pcANYWHERE host software, for example, the actual pcANYWHERE host program, AWHOST.EXE, could be specified with the following path notation:

```
C:\AWH\AWHOST.EXE
```

If you didn't add the AWH directory to your DOS search path, you would need to enter this entire command whenever you wanted to set up your host system to

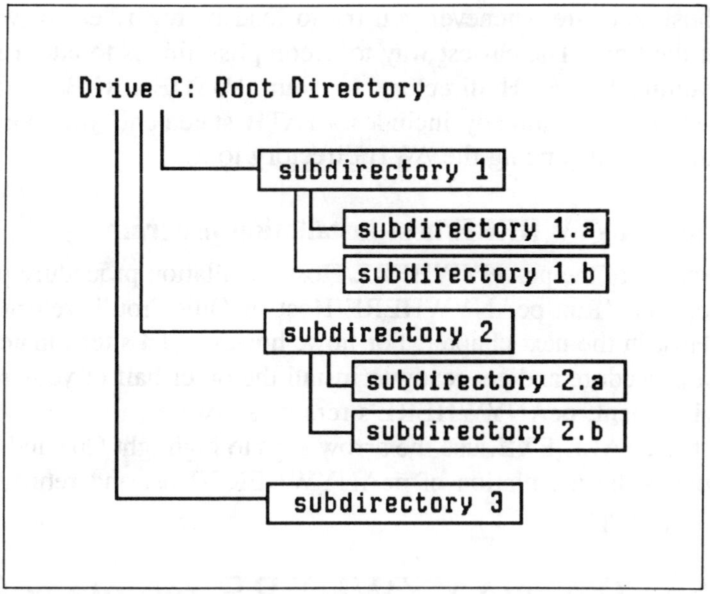

4-13 Setting up a search path helps DOS find and execute program files in a complex directory structure.

accept incoming calls. One alternative to this approach is to use the PATH command to specify a search path for your DOS environment. Understanding how and why search paths work requires knowing a little about what DOS does each time you enter a command.

When you press Enter after typing in a command, DOS follows a series of preordained steps in an attempt to execute that command. Specifically:

- DOS checks to see whether the command you issued is an internal DOS command—that is, a command that's loaded into RAM when you start your system, like COPY or DIR. If this is the case, DOS immediately executes that command.
- If the requested operation cannot be performed with an internal command, DOS tries to find the appropriate external command file in the currently active (default) directory. If this file is found, the command is executed.
- If the required file isn't found in the default directory, DOS still doesn't give up. Rather, it begins an organized search of any directories specified by a PATH statement, starting with the first directory listed. Once the appropriate file is found, DOS executes your command.
- If, after all this searching, DOS still can't find the required file, it displays a *Bad command or file name* message and returns you to the system prompt.

Because pcANYWHERE's AWH.EXE file is located in the C:\AWH directory, including this directory in your search path will allow DOS to find the pcANY-

WHERE host software whenever you try to load it, regardless of where you're working at the time. The easiest way to accomplish this is to add a PATH statement containing the AWH directory to your AUTOEXEC.BAT file. If your AUTOEXEC.BAT file already includes a PATH statement, you can accomplish the same thing by appending the AWH directory to it.

Exiting the pcANYWHERE/host installation program

The final screen of the pcANYWHERE/Host installation procedure provides you with two options, Run pcANYWHERE/Host or Quit. You'll return to pcANYWHERE/Host in the next chapter. For now, however, let's terminate the current installation procedure and get ready to install the other half of your remote-computing environment, pcANYWHERE's remote software.

To exit INSTALL.EXE, use the arrow keys to highlight Quit and press Enter. This completes the installation of pcANYWHERE/Host and returns you to the DOS system prompt.

Installing the pcANYWHERE remote software

This second portion of your pcANYWHERE installation will be quite a bit less complicated (and blissfully shorter) than the previous section. For one thing, much of the conceptual information I've already discussed applies equally to both the host and remote modules of pcANYWHERE. So there's no need to repeat it here. Second, all Windows-related pcANYWHERE functions are handled on the host side of the remote connection. Consequently, you won't have to cover those topics either. So let's get started, shall we?

INSTALL.EXE—the sequel

As was true of pcANYWHERE/Host, you must install the remote software using a utility named INSTALL.EXE. Although the two utilities share the same name, they're two separate programs. Which of your distribution disks contains the remote installation routine again depends on the type of disk drive you're using to install pcANYWHERE/Remote.

- If you're installing the remote software from a 5.25″ disk drive, INSTALL .EXE will be on a 5.25″ floppy disk marked Remote 1 Diskette.
- If you're installing the remote software from a 3.5″ disk drive, INSTALL .EXE will be on the 3.5″ microfloppy marked Remote Diskette.

Caution: As was true with the host software, you must run INSTALL.EXE (the remote version) to set up your remote PC. It, too, expands files originally compressed on the Remote Diskette(s).

To start INSTALL.EXE and begin installing pcANYWHERE/Remote on the computer you've selected as your remote system, place the appropriate remote diskette in drive A of the remote system. Then type A: and press Enter to make this the active disk drive. Finally, type INSTALL and press Enter.

After a few seconds, you'll see a screen resembling Fig. 4-14. This should look vaguely familiar, in that it varies only slightly from the initial screen of the pcANYWHERE/Host installation routine. The major difference is that the remote installation program lacks an option for installing pcANYWHERE to run with Windows 3. As I explained earlier, all Windows-related pcANYWHERE functions are handled on the host side of a remote-computing session.

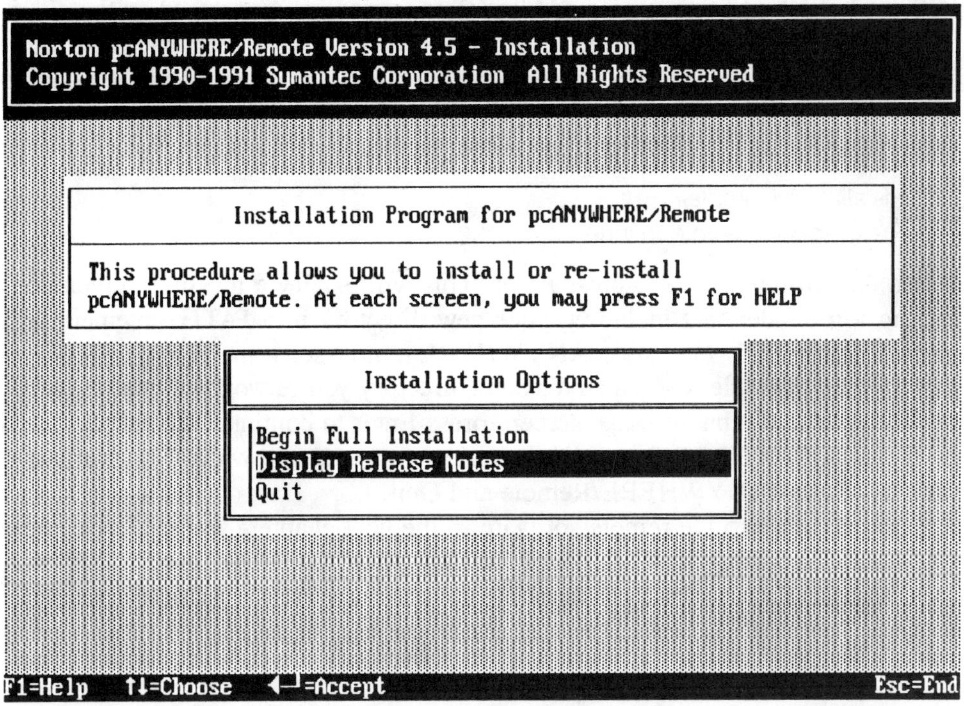

4-14 The pcANYWHERE/Remote installation process is almost identical to the pcANYWHERE/Host program installation.

To begin pcANYWHERE/Remote installation, use the arrow keys to highlight Begin Full Installation and press Enter. This will display another familiar screen, one which asks you to enter the full path name of the directory in which you want to install the pcANYWHERE remote program files. To specify a path for your pcANYWHERE/Remote files, type C:\AWR and press Enter.

Note: To specify a different drive or directory, simply substitute the appropriate information in the above step. For example, if you want the pcANYWHERE/Remote files installed on a floppy disk you've placed in drive B, as would be the case if you anticipate running the remote software on a dual-floppy laptop computer, you could specify B:\AWR.

Again, as happened when you were installing pcANYWHERE/Host, INSTALL.EXE requests that you enter a licensee name for the remote software. To identify a licensee, type in the name of the person or company authorized to use this copy of pcANYWHERE/Remote and press Enter. After collecting the above information, INSTALL.EXE will ask whether it should begin installing pcANYWHERE/Remote in the specified drive and directory. To begin the pcANYWHERE/Remote installation, use the arrow keys to highlight Yes and press Enter.

As was true during installation of pcANYWHERE/Host, the remote INSTALL.EXE utility begins expanding files originally compressed on the distribution disks and copying them to the specified disk and directory. (And once again, if you're installing the remote software from 5.25" disks, you'll be asked to insert the second distribution disk in drive A at some point during this process.) Once this process is completed, you'll see the following message:

Installation Complete
Press <Enter> to continue . . .

When this prompt appears, press Enter. This will display a message screen containing a reminder that including your new directory in a PATH statement will allow you to quickly load the pcANYWHERE remote software from the DOS system prompt, regardless of which disk and directory you're working in at the time.

After reading this message screen, press Enter to continue. The final screen of the pcANYWHERE/Remote installation procedure provides you with two options, Run pcANYWHERE/Remote and Quit.

You'll return to the remote software in the next chapter; you still have some fine-tuning to perform on both your host and remote systems. For now, however, let's close out this chapter by terminating the pcANYWHERE/Remote installation routine. To exit INSTALL.EXE, use the arrow keys to highlight Quit and press Enter.

Summary

This concludes the pcANYWHERE installation procedure. Both the host and remote programs have been transferred to their respective systems. To reiterate a previous suggestion, why not take this opportunity to store both of your pcANYWHERE distribution disks and the working disks you created earlier in a safe location? You never know when you might need them again. I'll meet you in the next chapter, where you'll start setting up the pcANYWHERE software to work properly with your host and remote systems.

5
CHAPTER

Configuring pcANYWHERE

With your hardware and software in place, it's time to start building a comprehensive telecommunications environment. The foundation of this structure, of course, will be pcANYWHERE's host and remote modules. So let's dig into each of these two powerful programs and see what's there.

Setting up a host environment

We'll begin on the receiving side of the remote connection, with pcANYWHERE/Host. Note: To be honest, my decision to begin with pcANYWHERE/Host corresponds to the flip of the coin that precedes most football games. In this particular case, the winning team elected to receive. I could just as easily have kicked things off with a discussion of pcANYWHERE/Remote. Regardless of which end of the pcANYWHERE field you start out with, however, everything will come together at the fifty-yard line once you make that first remote connection.

Starting the pcANYWHERE/Host program

There's nothing difficult about this task; to start pcANYWHERE/Host, type AWHOST and press Enter. Tip: This command assumes that you're either in the AWH directory or that you've added the directory to your DOS search path, as outlined in the previous chapter. If not, your job is only slightly more difficult, in that you'll need to use the DOS change directory (CD) command to make AWH the active directory before starting pcANYWHERE/Host.

This displays the pcANYWHERE/Host Main Menu, shown in Fig. 5-1. You'll see this initial menu screen whenever you enter the AWHOST command by itself to run the pcANYWHERE host software. Later on, you'll learn how pcANYWHERE provides several command line arguments that let you bypass its menu structure and initiate certain operations directly from the DOS system prompt. For now, however, let's take a look at what the various Host Main Menu options allow you to accomplish.

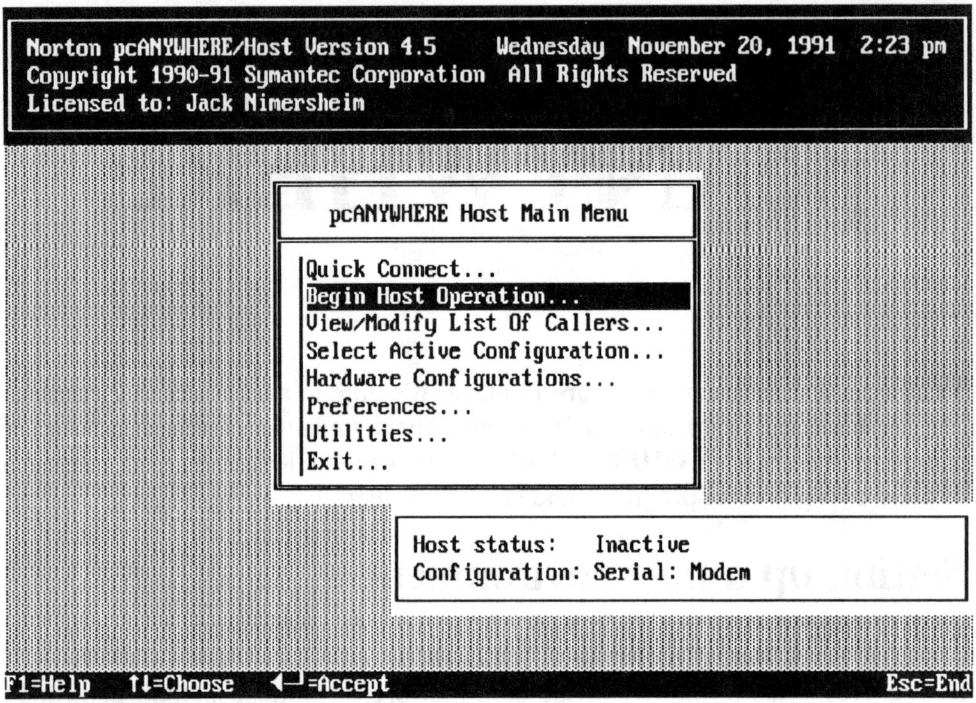

5-1 The pcANYWHERE/Host Main Menu.

The pcANYWHERE/Host Main Menu options

You might remember that, in an earlier chapter, I compared the host computer in a remote connection to the host of a party, inviting other systems to drop by for an electronic visit. With this analogy in mind, think of the Host Main Menu as a list of preparations any responsible host makes before his party actually begins.

For example, when planning a party most of us will put together an invitation list containing the names of those people we hope will attend. The View/Modify List Of Callers option on the pcANYWHERE Host Main Menu serves a similar purpose. It allows you to create and maintain a directory of authorized callers—that is, those individuals pcANYWHERE/Host will allow to take over your system from a remote location. As is true for an invitation-only party, you can set up your pcANYWHERE host environment so that persons not on this list will be turned away at the door, so to speak, should they try to establish a remote connection with your PC.

Note: Security is a major concern whenever you open an electronic door to your computer. After all, doing so makes it accessible (theoretically, at least) to anyone with a modem connected to his PC. Let's be honest: You wouldn't invite just anybody to a party at your house, would you? Of course not. As you'll see shortly—and, to an even greater extent, once you reach chapter 12, *Creating a secure environment*—when it comes to your electronic castle, you don't have to. pcANYWHERE includes a number of security features designed to provide peace of mind to anyone setting up a host system for remote computing. This said, let's set aside the party metaphor for a while and examine what each of the pcANYWHERE Host Main Menu options allows you to accomplish.

Quick Connect As its name implies, the Quick Connect option lets you prepare for a remote call quickly. You might select Quick Connect, for example, if someone using pcANYWHERE/Remote were making a one-time connection with your host system. Under these circumstances, you wouldn't necessarily want to include that individual in your permanent list of authorized callers. Because Quick Connect lets you prepare your host system for a remote connection by specifying a limited number of critical parameters, rather than forcing you to create a new entry in your caller list, it's the obvious choice for such a situation.

Begin Host Operation This is the option you'll select after you've finished setting up host environment to allow remote communications. It displays a second-level menu, which you then use to instruct pcANYWHERE/Host to wait for incoming calls from a remote system. (I'll discuss the various Begin Host Operation submenu options in the next chapter.)

View/Modify List of Callers As mentioned earlier, you use the View/Modify List of Callers option to create and maintain a directory of people you want to have access to your host system—the electronic analog of a list of individuals who will be admitted to your invitation-only party. Among other things, this option allows you to assign a user name and password to each authorized caller. You can then instruct pcANYWHERE/Host to refuse access to a caller and disconnect the phone line unless he enters both of these values correctly. (This is one of pcANYWHERE's built-in security features, which I alluded to earlier.)

Select Active Configuration One of pcANYWHERE's more impressive features is its flexibility. You can use the same host software with a variety of hardware configurations. In order for someone to contact your system over a 2400-bps remote connection, for instance, pcANYWHERE/Host would need to be set up to work with a 2400-bps modem. There might be other times, however, when you want to use a standard serial cable to establish a connection at much higher speeds between your host system and another computer located in the same room. The Select Active Configuration options allow you to quickly choose the appropriate parameters for either of these conditions.

Hardware Configurations Use this option to create and name various pcANYWHERE/Host configurations, from which you can then select your active settings using the previous option.

Preferences As was true of Begin Host Operation, the Preferences option also takes you to a submenu. From there, you can specify characteristics that define your pcANYWHERE/Host environment. Available preferences fall into three categories—Features, General, and Security—and range from whether you want to provide mouse support for remote operations to how much time someone calling in from a remote system has to enter his password before the host system terminates the connection.

Utilities The pcANYWHERE package includes a number of utility programs that allow you to maintain and, if necessary, modify your remote-computing environment. Some of these are diagnostic in nature, such as several utilities used to identify and test the hardware connected to your PC. Others provide a means for analyzing your remote operations, including viewing or printing a report showing the online activities of any individuals who called your host system. There's even a way for advanced users to modify specific drivers supplied with pcANYWHERE so that someone using a terminal not otherwise supported by pcANYWHERE/Host can remotely access a PC.

Exit Use the Exit option to close down pcANYWHERE/Host and return to the DOS system prompt. (As if you really needed me to explain this one to you.)

Two additional items, displayed in a separate box just below the Host Main Menu options, deserve mention:

Host Status This field identifies whether anyone is currently communicating with the host over a remote connection. The Inactive status in Fig. 5-1 indicates that the host system is not involved in a host/remote connection. This field would specify Active if a remote system were currently accessing pcANYWHERE/Host.

Configuration This field contains the active hardware configuration—that is, the name you assigned to the communication parameters currently in effect using Hardware Configurations and Select Active Configuration, two options described earlier. (Figure 5-1 lists Serial:Modem, the communication parameters following installation of the pcANYWHERE host software.)

With the above, admittedly brief descriptions as background, let's begin using some of these options to set up pcANYWHERE so it will run properly on your host system.

Setting preferences

Start by using the Preferences option to assign some global attributes to your pcANYWHERE/Host environment—that is, operating characteristics of pcANY-WHERE/Host that will be in effect whenever you run AWHOST unless they're overridden by the active hardware configuration. You can select an option from the pcANYWHERE/Host menu in one of two ways:

- Type the first letter of the option you want to select and press Enter.
- Use the arrow keys to highlight the desired operation and press Enter.

When working in either pcANYWHERE/Host or pcANYWHERE/Remote, you can use one of these methods to select a pcANYWHERE function from almost any menu or option box. Go ahead and select the Preferences option now, using one of the following methods: Type P and press Enter or use the arrow keys to highlight Preferences on the pcANYWHERE Host Main Menu and press Enter.

This displays the Preferences submenu, shown in Fig. 5-2. Notice that the options listed on this submenu correspond to the three categories mentioned earlier:

- Features
- General
- Security

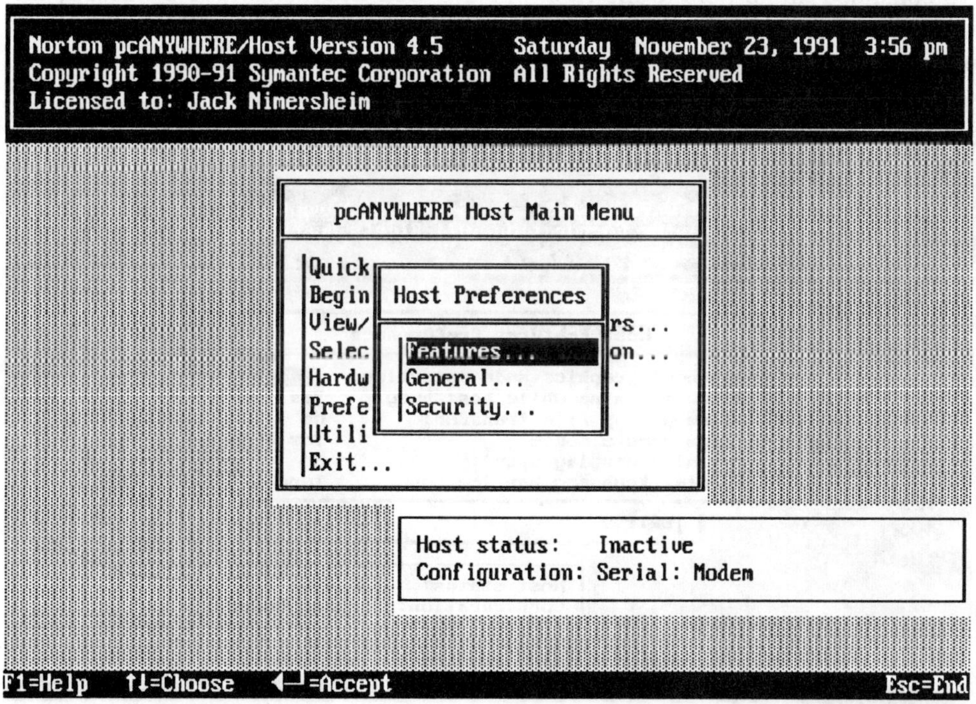

5-2 The Host Preferences menu.

This kind of multilevel structure—where selecting a menu item advances you to a secondary menu or, in some cases, the specific options available for a given activity—permeates virtually all pcANYWHERE operations. Consequently, using either pcANYWHERE/Host or pcANYWHERE/Remote is a simple matter of familiarizing yourself with the program's menu structure and then selecting the appropriate item to perform a particular task. (And remember, as pointed out in the previous chapter, on-line help is always a single keystroke away, the F1 function key, should you ever need to review how a specific menu or option is used.)

Setting Features Preferences

You'll begin customizing your host environment by examining the Features Preferences and even modifying one of them. To select the Features option from the Host Preferences menu, type F and press Enter or, if necessary, use the arrow keys to highlight the Features option and press Enter.

This displays the Host Features Preferences option box, shown in Fig. 5-3. Notice that, unlike a menu, this option box lists specific values associated with a number of pcANYWHERE program settings. Immediately following installation, for example, pcANYWHERE/Host is not set up to detect automatically whether or not an application program running on the host system is using advanced or nonstandard techniques to generate graphics-based displays, as indicated by the No contained in the field to the right of Advanced Graphics Mode Detection, the first item in the Host Features Preferences option box. Changing this field to Yes would modify pcANYWHERE/Host so that it would automatically detect nonstandard graphic displays.

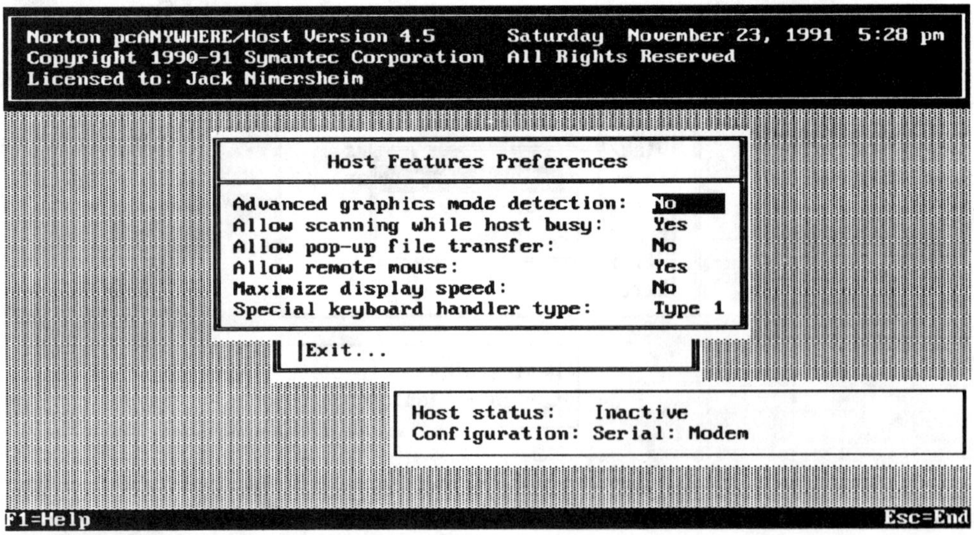

5-3 The Host Features Preferences option box.

As was true of the Preferences Menu, understanding how to change the settings in this particular option box involves learning techniques that you'll use elsewhere in pcANYWHERE/Host and pcANYWHERE/Remote. Before getting to that, however, let's examine briefly what the individual items in the Host Features Preference option box control:

Advanced Graphics Mode Detection As mentioned above, you use this option to make pcANYWHERE/Host compatible with programs that rely on nonstandard techniques to display graphic information. If all of the programs on your host system run properly during a remote connection, there's no need to change this option from its default No setting. If you discover that the displays of certain graphics-based applications appear garbled on the remote system, try changing this setting to Yes.

Allow Scanning while Host Busy Grammatically, this particular option might make your skin crawl. (It sounds like something out of an old Tarzan movie, doesn't it? "Me pcANYWHERE. Me scan while host busy.") Functionally, however, it can be a critical setting, especially for anyone using pcANYWHERE on a network. In some network environments, allowing pcANYWHERE to scan the display screen might interfere with normal network activity. If this happens, changing this setting to No will usually resolve the problem.

Allow Pop-Up File Transfer As you'll see later in this book, pcANYWHERE includes a special utility, AWSEND, which lets you quickly transfer files between the remote and host systems. You start AWSEND from the DOS system prompt—that is, when no other program is running on the host computer. In certain circumstances, however, it might be advantageous to let remote callers perform file transfers while they're actually working in an application on the host system. Setting this option to Yes allows callers to call up a special pop-up menu and perform file transfers in the background, as they continue to work in a foreground application. You should be aware of two things, however, if you plan to use this pcANYWHERE feature. First, background transfers are slower than using AWSEND and, second, enabling this feature adds approximately 18K to pcANYWHERE/Host's memory requirements.

Allow Remote Mouse This Preferences option, if enabled, also steals some RAM from the host system. In this case, however, it amounts to approximately 1K. That's why this option is set to Yes following installation. If none of the applications running on your host support mouse operations, changing this option to No will reclaim that 1K for other uses.

Maximize Display Speed Here's another memory-hungry option, one that requires approximately 46K of RAM when activated. It's also an option that affects only DOS (that is, nonWindows) applications running on the host. If you must rely on an extremely slow modem connection (300 or 1200 bps), you might need to sacrifice memory and change this option to Yes. Unless absolutely necessary, however, I strongly recommend you leave it set to No, the default setting.

Special Keyboard Handling Type This option is similar to Advanced graphics mode detection, which I described earlier. It allows you to modify pcANYWHERE to work properly with applications that use nonstandard techniques to process keyboard input. Three settings are available: Type 1, Type 2, and None. If an application running on the host ignores keystrokes entered at the remote system, you should try running pcANYWHERE/Host with this option set to either Type 1 (the default) or Type 2. However, you might not need any special keyboard handler, in which case you can set this value to None. That, in fact, is what I'm going to have you do in order to demonstrate how you modify fields in a pcANYWHERE option box.

To change this setting, use the arrow keys to highlight the Special Keyboard Handling Type option and press Enter. Then, at the subsequent option listing, press the up arrow key to highlight None and press Enter.

When you return to the Host Features Preferences option box, you'll notice that the last field now contains None, indicating that no special keyboard handlers will be enabled in your pcANYWHERE/Host environment. Now that this modification has been made, let's return to the Host Preferences menu.

To exit the Features option box, press Esc. pcANYWHERE recognizes that you've modified the current settings and gives you the opportunity to save this new configuration as your default. Therefore, these settings will be in effect each time you run pcANYWHERE/Host, as illustrated in Fig. 5-4. Go ahead and do this.

To save the new settings to your default configuration, with Yes highlighted, press Enter. This stores your new settings and returns you to the Host Preferences submenu. (See Fig. 5-2.)

Setting General Preferences

Next, continue customizing your pcANYWHERE/Host environment by modifying some of the General Preferences. To select the General option from the Host Preferences menu, type G and press Enter or, if necessary, use the arrow keys to highlight the General option and press Enter.

This displays the Host General Preferences option box, shown in Fig. 5-5. Once again, use this screen to specify global attributes of your pcANYWHERE/Host environment. Items specified in the Host General Preferences include:

Your Computer Name Assigning names to the various computers involved in remote/host connections allows pcANYWHERE to keep a record of your remote activities.

Eliminate Snow You'll need to worry about this setting only if any of your host or remote systems use a CGA display. In certain situations, CGA displays can gen-

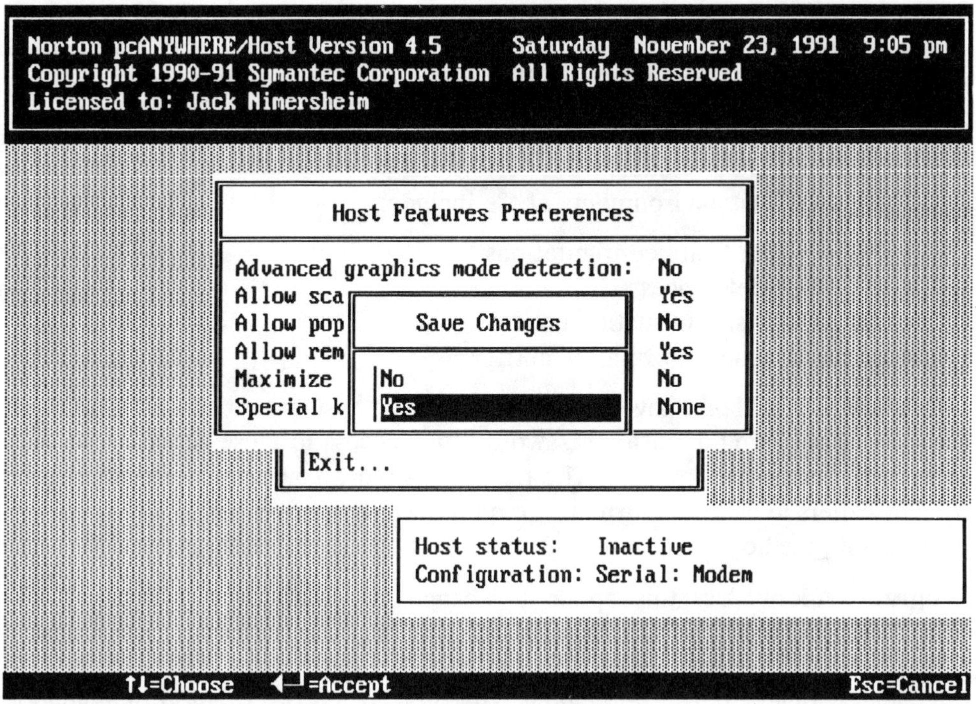

5-4 pcANYWHERE/Host asks you to verify that you want to save any changes made to your host environment.

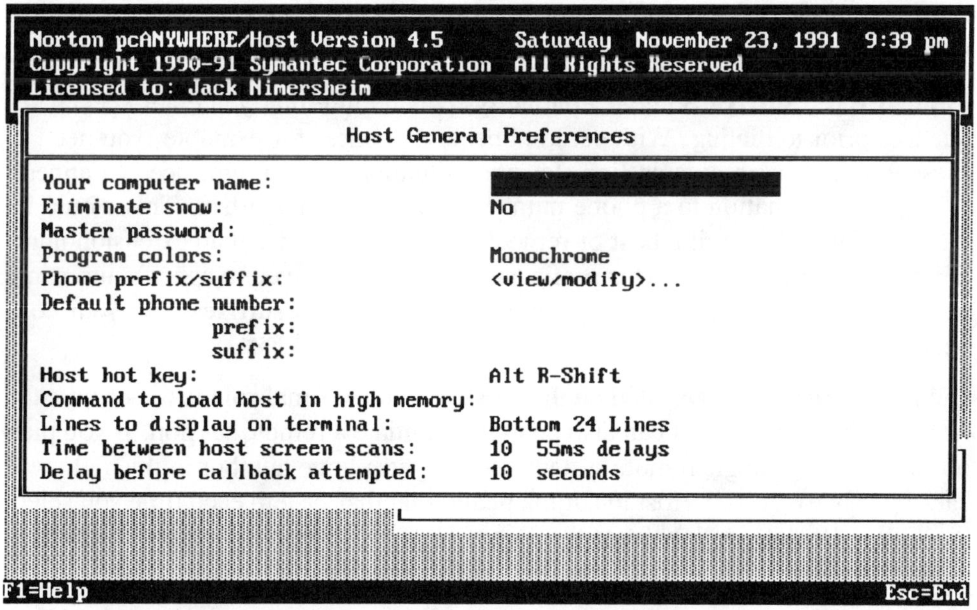

5-5 The Host General Preferences option box.

erate "snow," similar to the static on television sets. Setting this field to Yes can eliminate this problem should it arise. (You can use the Hardware Diagnostics option in the pcANYWHERE/Host Utilities submenu to test for snow.)

Master Password Assigning a master password to pcANYWHERE/Host prevents unauthorized individuals from performing certain activities associated with modifying your host environment. These include:

- changing hardware configurations
- changing preferences
- changing host information entries
- changing caller information entries

If this field is left blank, anyone with access to the host system can perform these activities. Specifying a master password limits access to these functions to only those individuals who know it. The master password you enter here can also be used by callers to override normal sign-on procedures and establish a remote connection with the host.

Program Colors Use this option to change the colors with which pcANYWHERE displays its menus, dialog boxes, messages, and the like. For the most part, you would use this option only to satisfy personal tastes. One exception to this general rule is if you're running pcANYWHERE on a laptop or notebook computer that uses an LCD screen. In this case, you would probably want to change the program color field from its original Default setting to Monochrome. (The figures in this book are also set to Monochrome in order to be reproduced in black and white.) The pcANYWHERE manual outlines the precise steps used to specify color combinations using the third option, User Defined.

Phone Prefix/Suffix Some telephone systems require that you enter special sequences prior to dialing. With certain business phones, for example, you need to press 9 to access an outside line. In other situations, you might need to append additional information to a phone number in the form of a suffix. This would be the case, for example, if a host or remote computer is assigned an extension number on a multiline system. This prefix/suffix field allows you to specify and name up to ten strings, which you can then use as prefixes and suffixes with your connect numbers.

Default Phone Number Although a host system's normal role is to accept calls from another computer, it can also be used to initiate a remote session. If you plan on calling only a single remote system, entering its phone number in this field will cause pcANYWHERE/Host to automatically dial that system each time you select the Call a Remote option from the Begin Host Operation submenu. Leave this field blank if you'll be calling multiple remote systems. Doing so causes pcANY-WHERE/Host to prompt you for a phone number before dialing.

Host Hot Key This field contains the key combination used to display the Host menu during a remote session. The default value for the hot key is an Alt–Right Shift key combination. You'll need to change the default setting only if it conflicts with a program that will be run on the host system. Choosing this option causes pcANYWHERE/Host to display a list of available hot-key combinations.

Command to Load Host in High Memory This option allows you to load a portion of the pcANYWHERE/Host program code into high RAM—that is, memory located between 640K and the one-megabyte memory address at which extended memory begins. To use this feature, your host system must use a memory-management utility such as Quarterdeck's QEMM-386 or DOS 5's EMM386.EXE to make this high RAM available for terminate-and-stay-resident (TSR) programs like pcANYWHERE/Host. If the host system includes these features, use this field to identify the command your memory manager uses to locate TSRs in high RAM. (With QEMM, for example, this command would be LOADHI. The comparable command for DOS 5's EMM386.EXE program is COMMAND /C LH.)

Lines to Display on Terminal If you're using pcANYWHERE only for PC-to-PC connections, you won't need to worry about this line. PC displays contain 25 display lines. If, however, your remote-computing environment includes remote terminals that include only 24 lines on their display, you can use this field to identify which portion of the host display should initially be shown on the remote terminal, the top or bottom 24 lines.

Time between Host Screen Scans During a remote connection, pcANY-WHERE/Host constantly scans the host display to determine whether any new information exists that needs to be transmitted to the remote system. If the host display flickers or appears jerky during a remote session, entering a lower number in this field will reduce the amount of time between each display scan. Be aware, however, that doing so also slows down operations on your host computer.

Delay before Callback Attempted As an added security precaution, pcANY-WHERE allows you to instruct your host system to call back selected remote systems rather than connecting them directly to the host computer. (This callback feature lets you verify that a specific remote system is being used to contact your host computer.) The number in this field indicates how many seconds pcANY-WHERE/Host will wait following disconnection before beginning to dial the number of a remote computer identified as a callback system.

You can leave most of the General Preferences settings unchanged. The default values assigned during the installation of pcANYWHERE/Host will work just fine on almost all systems. You should, however, assign a name and master password to your host computer. So, let's do that now. To assign a name and master password to your host system:

1. If necessary, use the arrow keys to highlight the Your Computer Name option and press Enter. Type MYHOST and press Enter.
2. Press the down arrow key once to highlight the Master Password option and press Enter.
3. Type in the master password you want assigned to your pcANYWHERE/Host environment and press Enter.

Notice that, after you complete the above steps, the Master Password field is filled with asterisks. pcANYWHERE/Host does this so that other people cannot use the General Preferences option box to see the master password you've assigned to your host system.

Tip: Try to make your master password something that's not obvious, but still easy to remember. Few experiences are as embarrassing as forgetting a password. Doing so also will force you to completely reinstall and reconfigure pcANYWHERE/Host because you'll no longer be able to perform several critical operations on the current installation.

With these changes entered, save your new configuration and return to the Host Preferences menu. To exit the General Preferences option box, press Esc. Once again, pcANYWHERE recognizes that you've modified the pcANYWHERE/Host environment and asks if you want to save this new configuration as your default settings.

To save the new General settings to your default configuration, with Yes highlighted, press Enter. This stores the new General settings and returns you to the Host Preferences submenu. (See Fig. 5-2.)

Setting Security Preferences

The third and final option on the Host Preferences submenu is Securities. Use this option to display an option box containing several security-related settings for your pcANYWHERE/Host environment. Rather than examine the Security Preferences now, however, I'm going to defer that discussion until chapter 12, *Creating a secure environment*, where I'll cover a wide range of topics pertaining to pcANYWHERE's ability to ensure the security of your remote operations. For now, let's exit the Preferences submenu and return to the Host Main Menu.

To exit the Preferences submenu, press Esc. This closes down the Preferences submenu and returns you to the Host Main Menu (see Fig. 5-1).

Specifying a hardware configuration

Now that you've assigned some global attributes to your pcANYWHERE/Host environment, let's use the Hardware Configurations option to begin defining the specific settings that will be used for your various on-line activities. Type H and press Enter, or use the arrow keys to highlight the Hardware Configurations option and press Enter.

Selecting the Hardware Configurations option displays a list containing the names you've assigned to the different settings available on your host computer. During installation, pcANYWHERE creates an initial configuration using certain default settings that it names Serial:Modem, as illustrated in Fig. 5-6. To get an idea of how hardware configurations work, let's examine the information this default configuration contains.

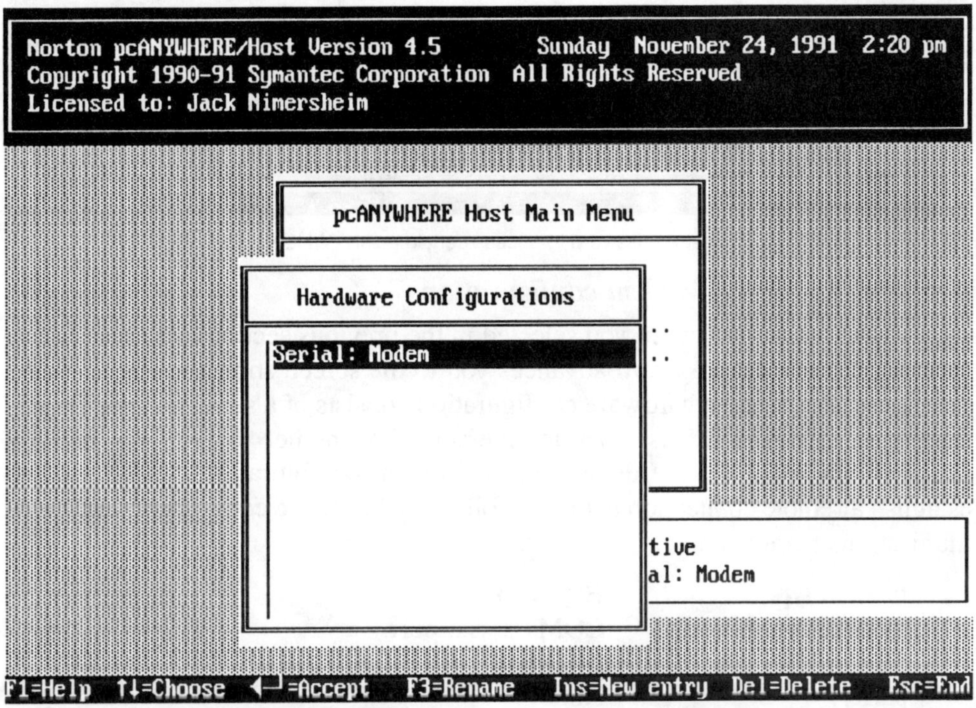

5-6 pcANYWHERE creates a default configuration, Serial:Modem, during installation.

Selecting the Serial:Modem configuration

Press Enter. Before granting you access to the Serial:Modem configuration screen, pcANYWHERE/Host displays the message shown in Fig. 5-7. This is the first time you've seen a pcANYWHERE security feature in action. Because you specified a master password in the previous section, pcANYWHERE requires that you enter this information before it will grant you permission to view and modify a hardware configuration.

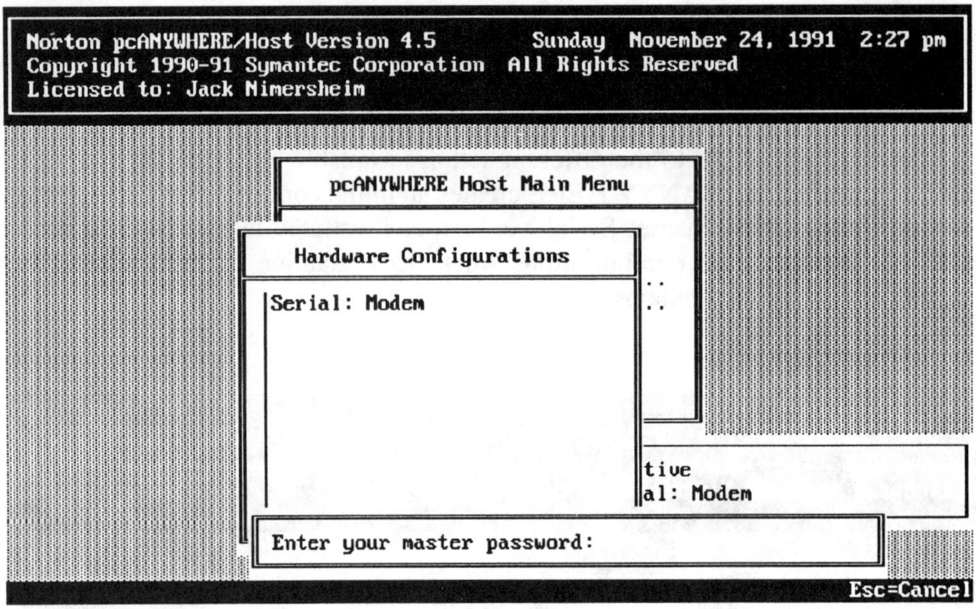

5-7 You must enter your master password to access certain pcANYWHERE/Host operations.

Accessing the Serial:Modem configuration

Type in the master password you selected in the previous section and press Enter. Entering the correct password advances you to the screen shown in Fig. 5-8. As this figure illustrates, a hardware configuration consists of a series of specific settings pcANYWHERE/Host uses to interact with another system. As initially installed, for example, the default Serial:Modem configuration can be used to establish a remote connection using the following hardware components and communications parameters:

modem type:	Hayes compatible
modem connection:	COM1 serial port
data-transfer speed:	2400 bps
parity:	None
DTR state:	on
RTS state:	on
phone system:	Touch Tone

Tip: If you're not sure what some of these terms mean, you might want to review the general information on telecommunications presented in chapter 2.

If this configuration matches your system parameters, there's no reason to change any of the Serial:Modem settings. But what if, for example, your modem is attached to COM2—or COM3, as it is on my system? If this were the case, you'd need to modify pcANYWHERE/Host's default settings before your host system could respond to an incoming call.

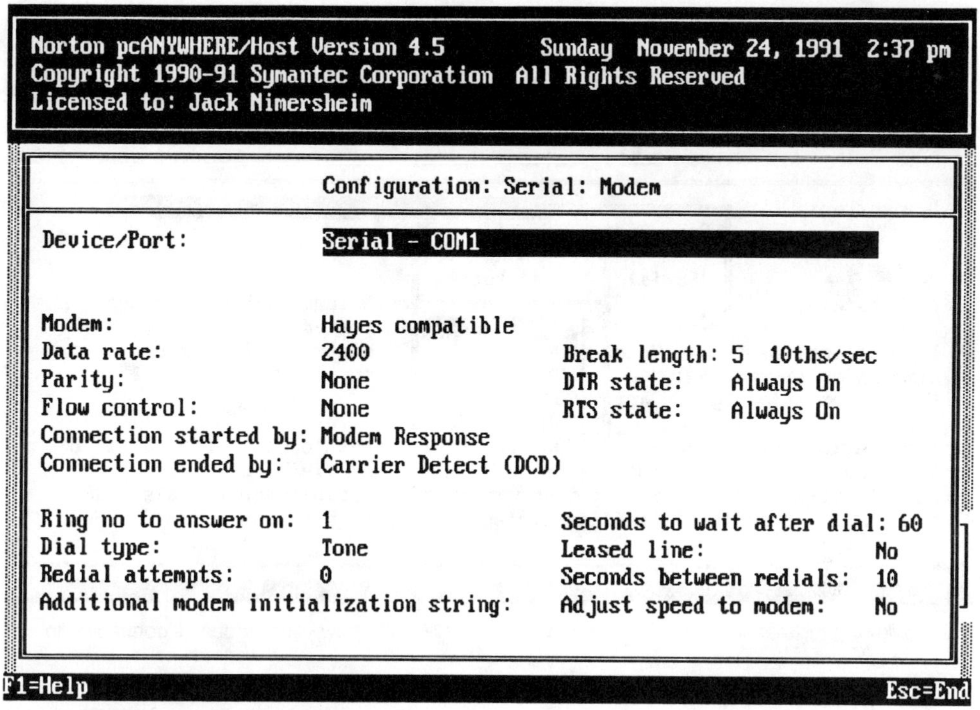

5-8 A Hardware Configuration option box.

Modifying the COM port setting

If necessary, use the arrow keys to highlight the Device/Port option and press Enter. At the Comm Device option box, press Enter to accept Serial. (The second option, Int14, is designed to be used with software that relies on special interrupt BIOS calls to communicate with a modem.) This displays the option box shown in Fig. 5-9. This option box identifies the serial (or COM) port to which your modem is attached.

To select a serial port other than COM1, use the arrow keys to highlight the serial port you want and press Enter. Whichever serial port you select will now be listed in the Device/Port field in the Serial:Modem configuration screen. While you're here, you might as well go ahead and change some other fields that could affect the performance of your host system running under the default configuration. Specifically, let's select the specific modem you own, if necessary, and specify a different data rate.

To select a different modem, highlight the Modem field and press Enter. Selecting the Modem field displays a box containing two options, as shown in Fig. 5-10. You can use the first option, None, to create a hardware configuration that supports direct communications between two computers in the same location—that is, a pcANYWHERE session in which a null cable, rather than a modem, is

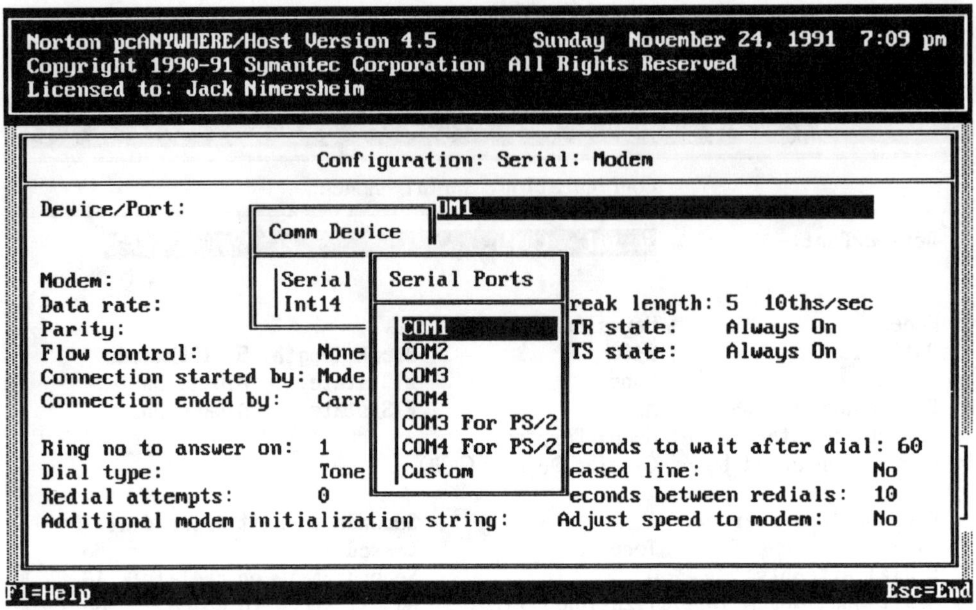

5-9 Following installation, you might need to tell pcANYWHERE how your modem is connected to your PC.

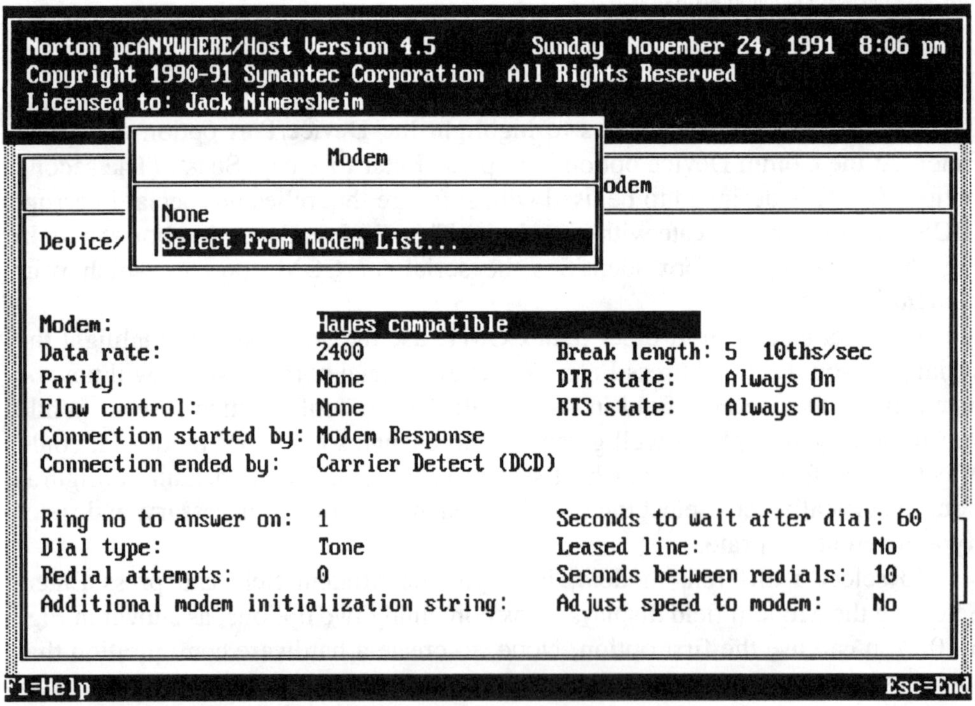

5-10 pcANYWHERE might support the specific type of modem you own.

used to connect the serial ports of the two systems involved. Use the second option to see if your modem is contained in the list of over 80 modems supported by pcANYWHERE/Host.

Viewing pcANYWHERE's modem list

If necessary, use the arrow keys to highlight the Select from Modem List option and press Enter.

This displays an option box containing an alphabetical listing of the various modems supported by pcANYWHERE. Using the PgUp, PgDn, and arrow keys, you can scroll through this option box and see if your modem is listed.

If it is, highlight its name and press Enter. If your modem is not supported by pcANYWHERE/Host, press Esc to leave the Modem field set to Hayes Compatible.

Tip: This assumes, of course, that your modem follows the Hayes standard, as most PC modems do. If you're using a nonHayes-compatible modem that isn't supported by pcANYWHERE, you'll need to select None from the Modem Options box and enter any commands needed to control your modem directly from the keyboard. Or, better yet, why not bite the bullet and buy a supported or Hayes-compatible model. I guarantee your life will be a lot easier in the long run.

The final setting you'll modify during this initial session is the Data Rate field, which is used to specify how quickly data will flow between the host and remote systems during a pcANYWHERE session.

Modifying the data transfer speed

Highlight the Data Rate option and press Enter. This displays the option box shown in Fig. 5-11, listing the various modem speeds available in pcANYWHERE. If you're using pcANYWHERE on both your host and remote systems, this value should be set to Max. If you're using a program other than pcANYWHERE on the remote system, set this value for the highest speed supported by the slower of the two modems involved in a host/modem connection. I'll assume here that you plan to run pcANYWHERE on both the host and remote systems.

Selecting the Max option

Use the arrow keys to highlight Max and press Enter. Note: The situation is a little more complicated if you'll be hooking up your host and remote systems over a direct connection. If this is the case, you'll need to know the fastest speed at which serial communications can occur on both systems. You should then set the Data Rate field to match the slower of the two.

The changes you've just made should allow pcANYWHERE/Host to work with almost any PC. As is true of most general statements, however, this one doesn't cover all possible situations. Before moving on to the pcANYWHERE/Remote software, therefore, I'll list a few conditions under which you'll need to modify additional fields in the default Serial:Modem configuration.

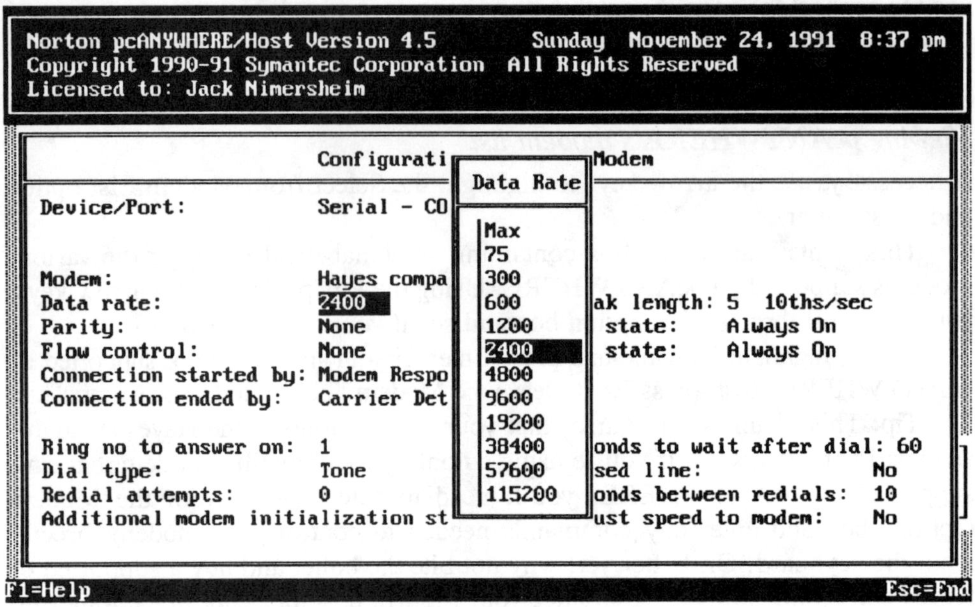

5-11 Selecting the appropriate data rate can improve the efficiency of your remote activities.

- Some modems that operate at speeds of 9600 bps or higher rely on some type of flow control to operate properly, usually a technique called RTS/CTS. If you're using a high-speed modem that requires this, change the Flow Control setting to RTS/CTS.
- Some PBX phone systems require you to change the Connection Started By setting to Ring Indicator (RI).
- If your local phone system does not support touch-tone dialing, a rarity in these modern times, you'll need to change the Dial Type setting to Pulse.

Because the number of different combinations of hardware and software that can be used for remote computing is almost infinite, it's impossible to cover all of them. For the vast majority of IBM-compatible PCs and modems, however, the above procedures should suffice to let you begin using your host system. But before actually doing so, you still need to make sure your pcANYWHERE/Remote software is set up properly to run on your remote system. To do that, return to the Host Main Menu and exit pcANYWHERE/Host.

To return to the Host Main Menu, press Esc to exit the Serial:Modem configuration screen. Press Enter at the Save Changes prompt to store the new Serial:Modem settings as your default configuration. Press Esc to exit Hardware Configurations.

This will return you to the pcANYWHERE Host Main Menu. Don't worry, you'll be returning to pcANYWHERE/Host in the next chapter. For now, however, move over to the remote side of your pcANYWHERE connection.

To exit pcANYWHERE/Host, type E and press Enter or use the arrow keys to highlight the Exit option and press Enter. At the Exit AWHOST prompt, press Enter to specify Yes. This will close down pcANYWHERE/Host and take you back to the DOS system prompt. Now let's move over to your remote system and make sure it's set up properly to run pcANYWHERE.

Setting up a remote environment

There's nothing to stop you from installing pcANYWHERE's host and remote software on the same PC. If you want to establish remote connections, however, they'll need to be running on two separate computers. I'm going to assume, therefore, that the latter is the case and have you set up pcANYWHERE/Remote as if it were going to be used to call up and take over the pcANYWHERE/Host software configured in the previous section. So get to your remote computer, wherever it's located, and let's get started.

Starting the pcANYWHERE remote program

As was true of pcANYWHERE/Host, this one's simple; to start pcANYWHERE/Remote, type AWREMOTE and press Enter. Note: Once again, I'm assuming that you're either in the appropriate directory or that you've added this directory to your DOS search path, as outlined in the previous chapter. If not, as was true for pcANYWHERE/Host, you'll need to use the DOS change directory (CD) command to make AWR the active directory before starting pcANYWHERE/Remote.

This loads the pcANYWHERE remote software and displays its Main Menu, shown in Fig. 5-12. Several of the items on this menu should look familiar, because they also appeared in pcANYWHERE/Host.

The pcANYWHERE/Remote Main Menu options

As you did with pcANYWHERE/Host, let's begin by examining briefly how the different pcANYWHERE/Remote Main Menu options are used.

Quick Connect As was true in pcANYWHERE/Host, the Quick Connect option lets you place a remote call to a host system not recorded in your list of hosts. In essence, the pcANYWHERE/Remote Quick Connect option transforms your remote system into a generic terminal. In other words, it sets up direct communications between you and your modem, allowing you to perform a wide range of telecommunications-related activities.

Connect Via List of Hosts This is one of the menu items pcANYWHERE/Remote does not share with its host-based counterpart. pcANYWHERE assumes that most remote computing sessions will originate at the remote system—that is, that someone sitting at a remote system has called a computer running pcANYWHERE/Host. It's a logical assumption. To simplify matters, pcANYWHERE/

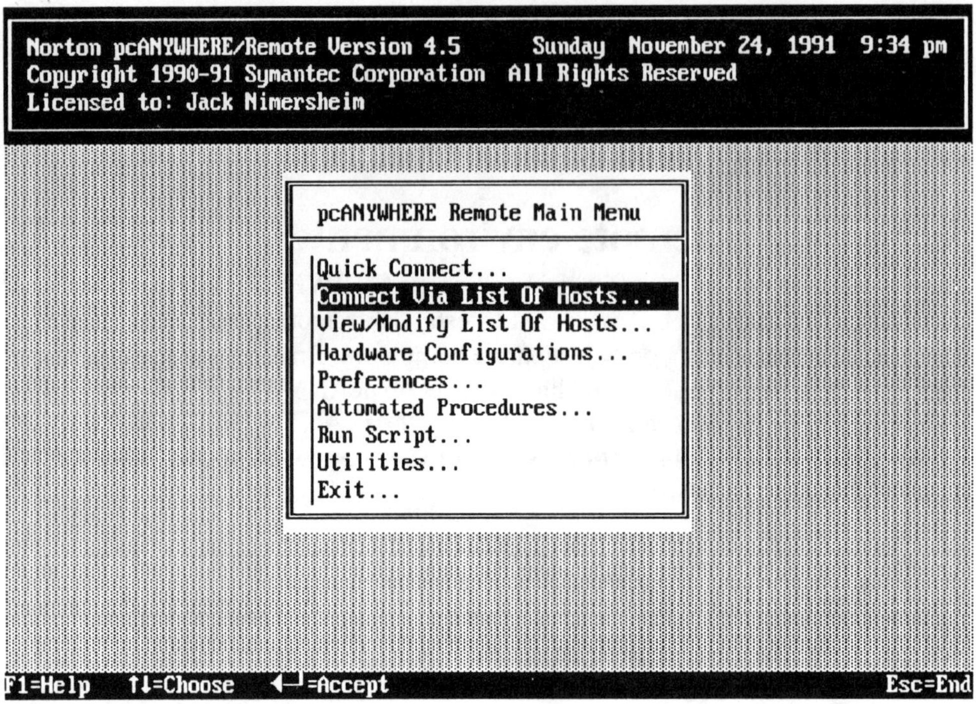

5-12 The pcANYWHERE/Remote Main Menu.

Remote allows you to create a host directory, in which you record specific information for each of the various host systems you anticipate calling. Once this directory (called a *list of hosts*) exists, you can use this option to automate the process of contacting a specific host system.

View/Modify List of Hosts As was true in a similar option in pcANYWHERE/Host, you use this option to create and maintain a directory of the host system you expect to call regularly. Among other things, this option allows you to identify a host's name, telephone number, password, and the hardware configuration it uses for a remote session. Once a host system is recorded in the list of hosts, you can use the previous option to automate the process of contacting it.

Hardware Configurations Use this option to define different kinds of hardware configurations, which can be assigned to the various host systems recorded in your list of hosts.

Preferences As was true in pcANYWHERE/Host, you use the pcANYWHERE/Remote Preferences option to specify the characteristics of your remote environment. The pcANYWHERE/Remote Preferences submenu contains five options, or categories of preferences: File Transfer Protocol, General, Mouse, Remote Access Session, and Terminal Emulation Session.

Automated Procedures As its name implies, this option allows you to define procedures that can then be used to fully automate your online activities. This is an extremely advanced and exceptionally useful feature of pcANYWHERE—one I'll be discussing in detail in chapter 11, *Automating your on-line activities*.

Run Script Beginning with version 4.5, Symantec has added a powerful script language to pcANYWHERE. Like procedures, scripts can be used to automate complex online activities.

Utilities During installation, pcANYWHERE/Remote places a number of utility programs in your AWR directory. As was true with pcANYWHERE/Host, these include diagnostic utilities and several methods for reviewing your online activities. Unique to pcANYWHERE/Remote is a special utility that lets you test scripts to see if they contain any syntax errors, incorrectly written commands that could cause a script to malfunction during execution.

Exit Use the Exit option to close down pcANYWHERE/Remote and return to the DOS system prompt.

As you did with pcANYWHERE/Host, you'll begin by using some of these options to set up pcANYWHERE/Remote to run properly on your remote system.

Setting Preferences

Once again, start by choosing the Preferences option and assigning some general attributes to your remote environment. To select the pcANYWHERE/Remote Preferences option, type P and press Enter, or use the arrow keys to highlight Preferences on the pcANYWHERE Host Main Menu and press Enter.

This will display the pcANYWHERE/Remote Preferences submenu, shown in Fig. 5-13, which contains the following suboptions:

- File Transfer Protocol
- General
- Mouse
- Remote Access Session
- Terminal Emulation Session

You'll notice that, unlike pcANYWHERE/Host, the pcANYWHERE/Remote Preferences submenu contains no Features option. That's because, of necessity, most operational attributes affecting remote connections must be defined at the host end of a remote/host connection. It is, after all, the system that turns control of itself over to another computer. So, let's begin customization of the pcANYWHERE/Remote environment with the general preferences.

To select the General option from the Remote Preferences menu, type G and press Enter or, if necessary, use the arrow keys to highlight the General option and press Enter.

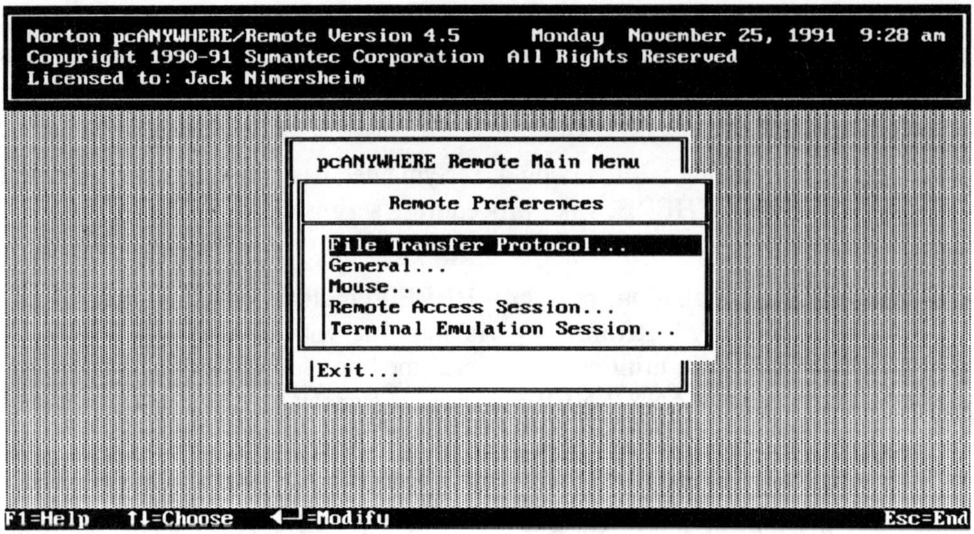

5-13 The Remote Preferences menu.

This will display the Remote General Preferences option box, shown in Fig. 5-14. As was true in pcANYWHERE/Host, you use this screen to specify global attributes of your remote environment. Items specified in the Remote General Preferences box include:

Your Computer Name Assigning names to your remote system allows pcANYWHERE to keep a record of your remote activities.

Eliminate Snow Here, again, you'll need to worry about this setting only if any of your host or remote systems uses a CGA display. (pcANYWHERE/Remote's Utilities menu also contains an option that tests for snow.)

Master Password As was true with pcANYWHERE/Host, specifying a master password for your remote environment prevents unauthorized individuals from accessing certain pcANYWHERE/Remote functions. Leaving this field blank permits anyone with access to your computer to select any option from the various pcANYWHERE/Remote menus.

Program Colors Use this option to change the colors pcANYWHERE/Remote uses to display its menus, dialog boxes, messages, and the like. Again, you might want to select the Monochrome option if you're running pcANYWHERE/Remote on a laptop or notebook computer that uses an LCD screen.

Phone Prefix/Suffix If the phone system you're using with pcANYWHERE/Remote requires that you enter special dialing (as I explained in the Host section, for example, for PBX systems), you can use this option to identify them. Like pcANYWHERE/Host, pcANYWHERE/Remote lets you specify and name up to

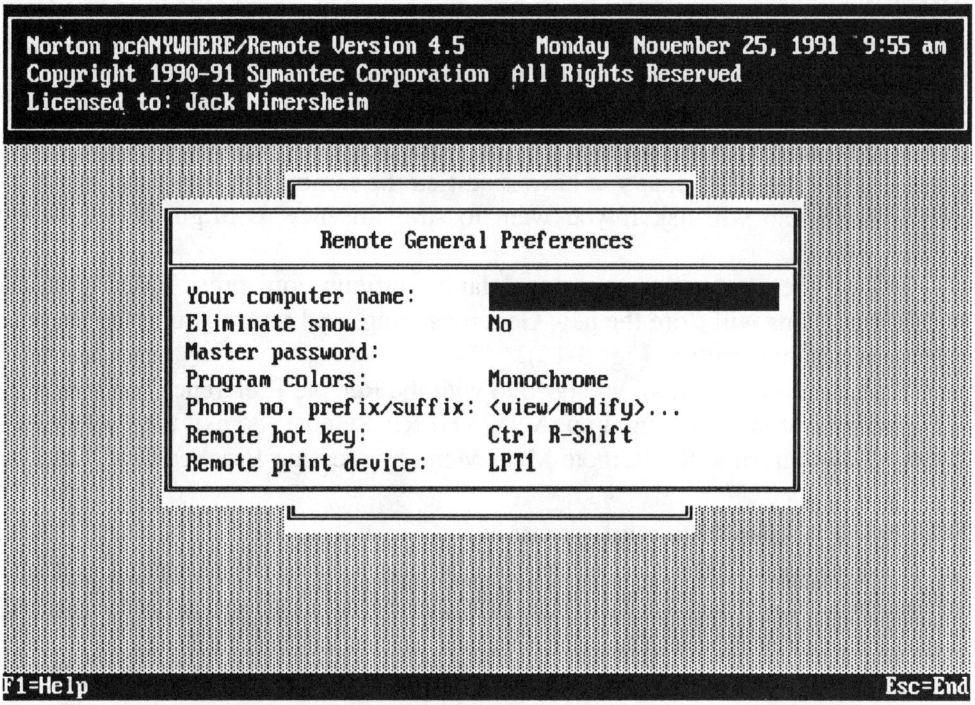

5-14 The Remote General Preferences option box.

ten strings, which you can then use as prefixes and suffixes with your connect numbers.

Remote Hot Key This field contains the key combination used to display the Session Options menu during a remote connection. The default value for the remote system is a Ctrl—Right Shift key combination. Once again, you'll need to change this default setting only if it conflicts with a program that will be run on the host system.

Remote Print Device Use this field to specify how print operations initiated by applications running on the host system will be performed at the remote site. (pcANYWHERE/Remote ignores this setting if the host system is not configured to support remote printing.)

As you did in the previous section, limit your initial activities in General Preferences to assigning a name and master password to your remote system. Use the arrow keys to highlight the Your Computer Name option and press Enter. Type MYREMOTE and press Enter.

Press the down arrow key once to highlight the Master Password option and press Enter. Type in the master password you want assigned to your pcANYWHERE/Remote system and press Enter. As it did on the host system, pcANY-

WHERE will protect your master password from prying eyes by filling this field with asterisks.

Tip: Once again, select a master password that's not obvious, but easy to remember. With these changes recorded, save your new configuration and return to the Remote Preferences menu. To exit the General Preferences option box, press Esc. Recognizing that you have modified the remote environment, pcANYWHERE/Remote will ask if you want to save this new configuration as your default setting.

To save the new settings to your default configuration, press Enter with Yes highlighted. This will store the new General settings and return you to the Remote Preferences submenu (see Fig. 5-13).

You've accomplished all you need to with the Remote Preferences options during this initial stage of setting up pcANYWHERE/Remote. So exit the Preferences submenu and return to the Remote Main Menu by pressing Esc (see Fig. 5-12).

Specifying a hardware configuration

As we did with pcANYWHERE/Host, the next thing we'll do is use pcANYWHERE/Remote's Hardware Configurations option to begin defining specific settings that ultimately will be used for your various on-line activities.

To select the Hardware Configurations option, type H and press Enter or use the arrow keys to highlight the Hardware Configurations option and press Enter.

Notice that installing pcANYWHERE/Remote also created an initial configuration called Serial: Modem. So, let's select this default configuration and see whether it needs to be modified for your remote system.

To select the Serial: Modem configuration, press Enter. Because you specified a master password in the previous section, you'll need to enter that information before pcANYWHERE/Remote will grant you access to the Serial: Modem configuration screen.

To access the Serial: Modem configuration, type in the master password you selected in the previous section and press Enter. Entering the correct password advances you to the screen shown in Fig. 5-15. This certainly looks familiar, doesn't it? It should. The pcANYWHERE/Remote configuration screen is identical to the one used by pcANYWHERE/Host, except that it is used to specify the appropriate hardware settings and communications parameters for your remote computer.

As was true on the host system, there's no reason to change the Serial: Modem configuration if the original settings match the system parameters for your remote computer. If not, go ahead and modify them now, using the same procedures you did in the previous section to set up pcANYWHERE/Host to run properly on your host system. After doing so, you can return to the Remote Main Menu.

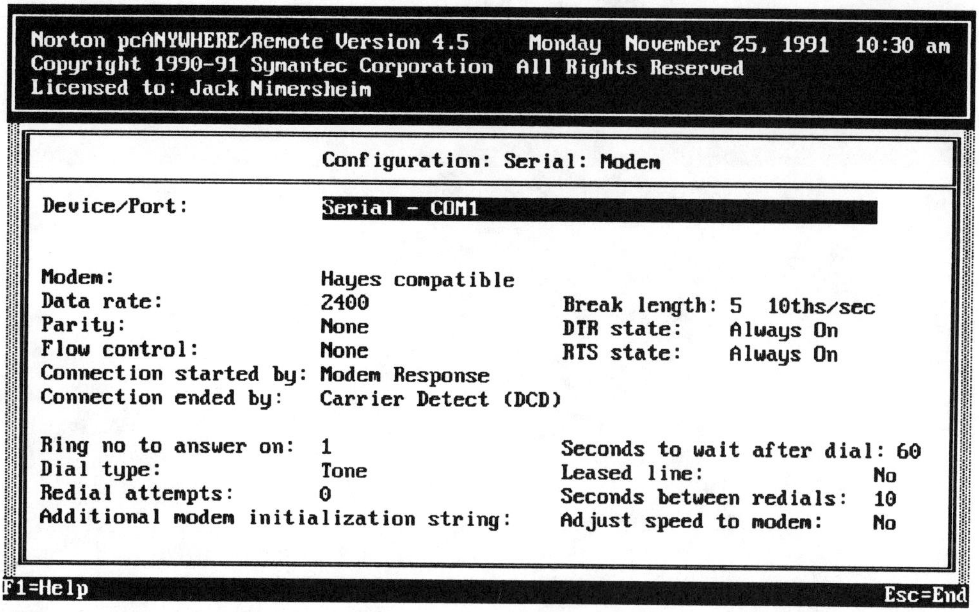

5-15 Remote systems must also be configured to work properly with your PC hardware.

To return to the Remote Main Menu, press Esc to exit the Serial: Modem configuration screen. Press Enter at the Save Changes prompt to store the new Serial: Modem settings as your default configuration. Press Esc to exit Hardware Configurations. This returns you to the pcANYWHERE Remote Main Menu. To close out this chapter, let's exit pcANYWHERE/Remote and return to the DOS system prompt.

To exit pcANYWHERE/Remote, type E and press Enter or use the arrow keys to highlight the Exit option and press Enter. At the Exit AWREMOTE prompt, press Enter to specify Yes. This closes down pcANYWHERE/Remote and takes you back to the DOS system prompt.

Summary

At this point, having customized both pcANYWHERE/Host and pcANYWHERE/Remote to run properly on their respective systems, you're ready to establish an initial connection between the two and actually begin remote computing. That's exactly what you'll do in the next chapter, which begins Part 2 of this book, *Using pcANYWHERE*.

6
CHAPTER

Using pcANYWHERE /Host

In this chapter you're going to concentrate on the host side of the pcANYWHERE remote-computing connection, i.e., the computer running pcANYWHERE/Host. You've already seen how several of the pcANYWHERE/Host menu options can be used to assign general attributes to your host environment—modem and COM port settings, display characteristics, password security, and the like. Having laid this groundwork, you can now begin creating individual listings for each caller authorized to access your host system over a remote connection. After doing this, I'll briefly discuss the various options available to you once you select the Begin Host Operation option from the pcANYWHERE/Host Main Menu. To start things rolling, you'll need to get back into the pcANYWHERE/Host software.

To run pcANYWHERE/Host, type AWHOST and press Enter on your host system. Note: Remember, this step assumes that you're either in the AWH directory or that you've added that directory to your DOS search path. If not, you'll need to type CD \ AWH and press Enter to make AWH the active directory before entering the above command.

As you saw in the previous chapter, starting the AWHOST program displays the Main Menu shown in Fig. 6-1. Before actually placing fingers to the keyboard, however, let's briefly discuss another useful security feature of pcANYWHERE/Host.

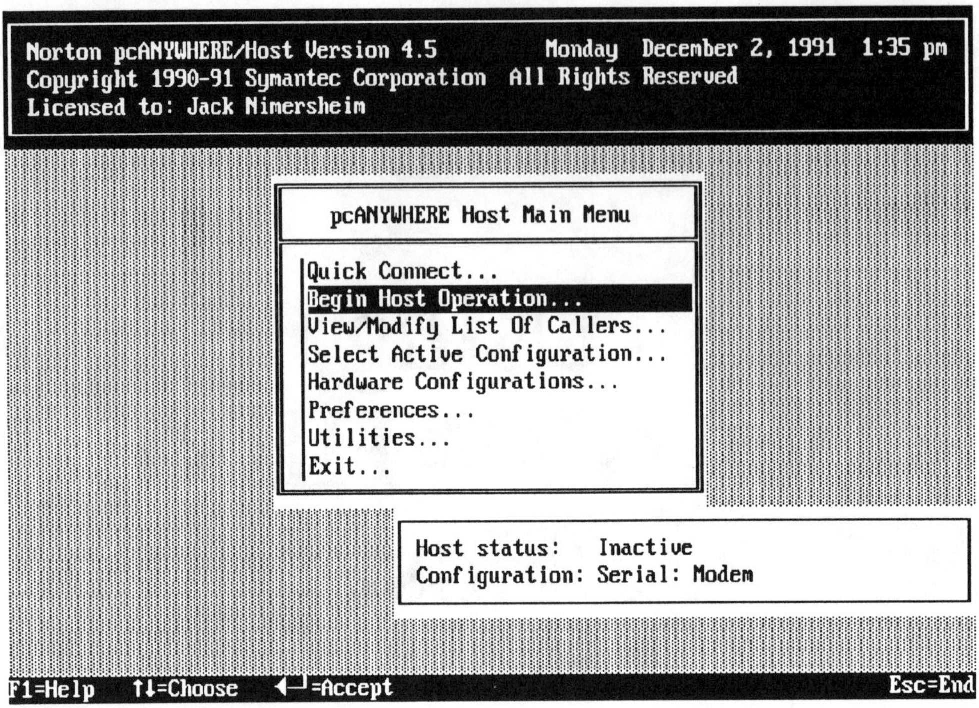

6-1 The pcANYWHERE/Host Main Menu.

Working with the caller list

You might recall that I earlier compared the host system in a pcANYWHERE session to the host of a party. And make no mistake about it, your on-line party is fairly exclusive. Someone's name must appear on an official "invitation list" before he can attend. In the pcANYWHERE scheme of things, this invitation list consists of a series of caller entries containing information about the people who you will allow to call your host system from a remote computer. Few people open their home to anyone who feels like dropping by without having made some kind of prior arrangements. The likely result of such a *laissez faire* attitude would be potential pandemonium. pcANYWHERE's caller list prevents similar chaos from occurring on your pcANYWHERE host system.

In addition to protecting your electronic home against unauthorized visitors, so to speak, information recorded within the various entries in the caller list defines how different individuals can interact with your host system during a remote-computing session. You can, for example, set up pcANYWHERE to let certain callers blank the display on the host system, a useful security measure for preventing prying eyes from observing potentially sensitive files that are being accessed remotely on your host PC. It's also possible to limit the length of time a caller can remain connected to your host system each time he contacts your host system. But rather than merely describing the caller list, let's see how it works.

Adding entries to the caller list

The View/Modify List of Callers option on the pcANYWHERE/Host Main Menu provides access to the list of callers authorized to contact your host system. To access your pcANYWHERE/Host caller list, type V and press Enter, or use the arrow keys to highlight View/Modify List of Callers on the pcANYWHERE Host Main Menu and press Enter.

Whenever you select this option, pcANYWHERE/Host displays the names of any individuals currently entered in your caller list. Immediately following the installation of pcANYWHERE, this list is empty, as illustrated in Fig. 6-2.

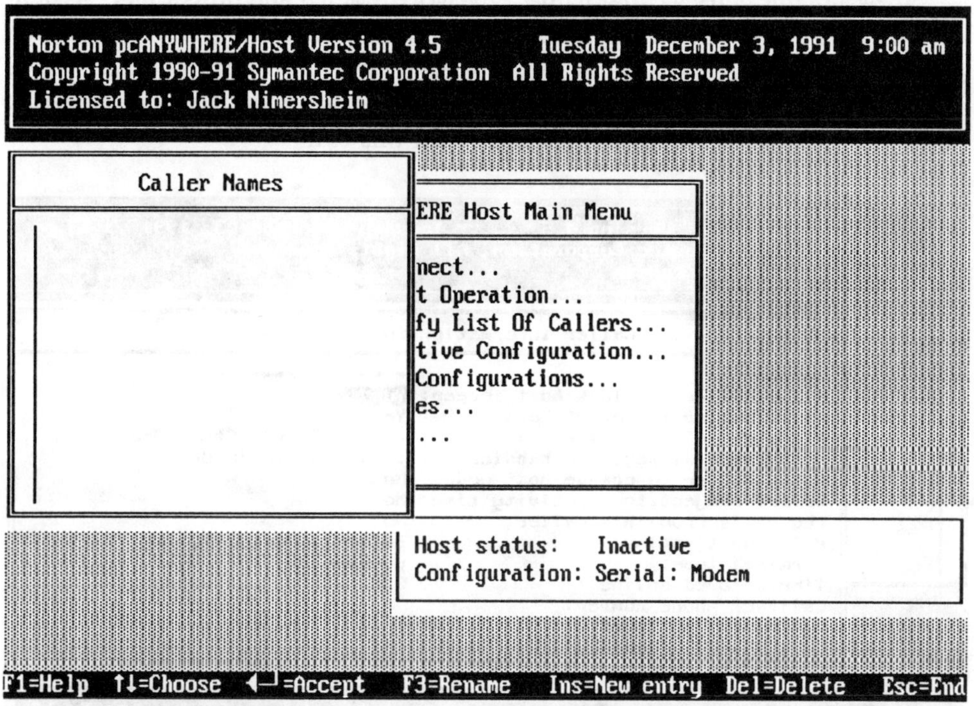

6-2 pcANYWHERE/Host lets you create a list of caller names representing individuals authorized to contact your host system.

As is true with many pcANYWHERE operations, the various options available to you at this time are listed across the bottom row of Fig. 6-2. Your old friend, Help, is there, as are instructions to use the arrow keys and Enter key to highlight and select entries. Of course, as I've already pointed out, our caller list is currently empty. So why don't you begin building one?

To add a new entry to your caller list, press Ins, the Insert key, to specify a new entry. Note: Because adding new entries to the caller list represents a modification to the host environment, pcANYWHERE won't allow you to continue past this point unless and until you enter the master password you assigned to your host system in the previous chapter. Go ahead and do this now.

Selecting the New Entry option displays a prompt box, requesting that you enter a name for your new caller entry. As you'll see shortly, pcANYWHERE/Host lets you create and maintain a log of your remote-access activities. Among other things, the name you enter here will be used to denote log entries for this individual. You should, therefore, pick names that allow you to easily identify each caller. Caller names, which can be up to 30 characters long, don't have to match the computer name you assigned to your remote computer in the previous chapter. To avoid confusion, however, you can use a variation of that name for your sample entry.

To assign this entry a caller name, type MY REMOTE and press Enter. Assigning a name to a new caller entry advances you to the Caller Information dialog box shown in Fig. 6-3. Use this dialog box to specify how this particular caller will be able to interact with pcANYWHERE/Host during a remote session. The first thing you're asked to specify is a password for this entry.

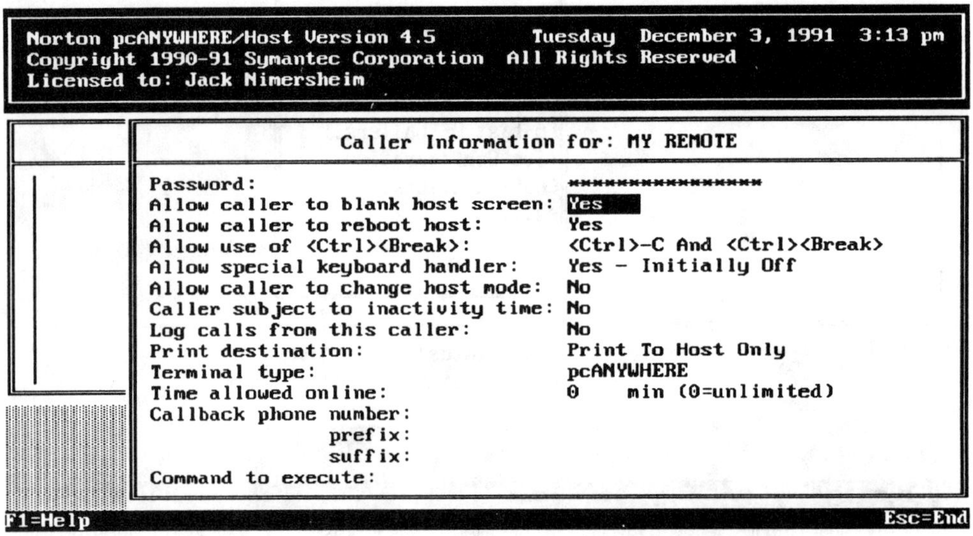

6-3 Each name on the caller list is associated with a Caller Information file.

To assign a password, type TEST CALL and press Enter. Notice that, when you press Enter, pcANYWHERE replaces the actual password you typed in with a series of asterisks (*). This prevents prying eyes from seeing the passwords you've selected that, in turn, grant callers access to your host system.

When adding new caller information, the only requirement is that you assign each entry a unique password—that is, a password not associated with any other caller entry. How many and which other fields shown in Fig. 6-3 you change is entirely your choice. Understanding how pcANYWHERE/Host uses caller passwords will help you decide which fields, if any, you want to modify for a particular entry in your caller list.

Whenever someone attempts to call your pcANYWHERE/Host, the first thing he's asked is to enter a password. Doing so triggers several responses on the host system:

- pcANYWHERE/Host compares the entered password to a master list of all valid passwords assigned to the various Caller Information files in your caller list.
- If no match is found, in other words, if the caller enters an invalid password, pcANYWHERE/Host denies access to the host system.
- If a match does exist, pcANYWHERE/Host grants the caller access to the host system and uses the settings in the Caller Information file associated with that password to determine what type of activities he'll be able to perform on your host system.

A valid password, therefore, corresponds to an invitation to the imaginary party. Persons without one aren't granted permission to enter your host system. Furthermore, the type of invitation a particular guest possesses governs what he has permission to do once he arrives. To clarify what I mean by this, let's refine this now-familiar party metaphor even further.

You can also use passwords to determine access privileges. Suppose, for example, that you were planning a party that consisted of two different events—say a sit-down dinner followed by a less formal dance. Additionally, let's assume that you weren't going to include everyone in both events. Some people would be asked to both the dinner and the dance, while others would be invited to the dance only. Which portion of the party a given guest could attend would depend on what type of invitation he had received.

The passwords in the various entries of your caller list serve much the same function. When you decide which password will be used by a specific caller, you effectively define the access privileges that person has once he's connected to your host system.

It's possible to assign the same password—and, therefore, equal access privileges—to more than one caller. Each Caller Information file, however, must be associated with a unique password.

The Caller Information fields

Each field in the Caller Information dialog box corresponds to a setting that, in turn, influences how someone signing on with the password assigned to that listing can interact with your host system during a remote-computing session. Values specified in the Caller Information dialog box include:

Allow Caller to Blank Host Screen

The default value for this field is Yes, meaning that the remote caller will be able to turn off, or disable, the display on the host computer. Changing this field to No stops a caller from blanking the host display, thus ensuring that you'll be able to keep tabs on what activities callers are performing on your host.

Allow Caller to Reboot Host

When set to the default Yes value, this option allows a remote caller to reboot the host computer. This ability could come in handy if, for example, the host system were to freeze up during a remote session. Changing the value in this field to No disables the Reboot Host Computer option on the pcANYWHERE/Remote menu.

Tip: For the reboot option to have any practical value, you'll need to set up the host system to automatically run pcANYWHERE/Host and prepare itself to accept incoming calls each time it's turned on or rebooted. The easiest way to accomplish this is to add the following two lines to the end of the AUTOEXEC .BAT file on the host system:

```
CD \AWH
AWHOST -m=w
```

Allow Use of CtrlBreak

The setting you choose for this field depends on how much control you want the remote caller to have over terminating DOS operations and application programs running on the host system. As you'll see in a moment, the Command to Execute field in the Caller Information screen can be used to run an automated procedure each time a person using this password signs on to pcANYWHERE/Host. (You could use this feature in a business setting to automatically transfer a specific file—a sales report, for example—to the remote system, after which you would end the session.) Theoretically, the caller could terminate your preprogrammed procedure prematurely by entering a Ctrl–Break or Ctrl–C key combination. Setting this field to No prevents a caller from doing so.

Allow Special Keyboard Handler

As explained in the previous chapter, pcANYWHERE provides several keyboard handlers designed to eliminate potential incompatibilities between the host and remote system. If necessary, you can use this field to specify whether a caller can invoke a keyboard handler during a remote session, and even specify which one.

Allow Caller to Change Host Mode

When set to No, the default setting, a caller has no control over how a remote session is terminated. Setting this field to Yes allows the caller to specify what the host computer should do when a remote session ends. Options available to the caller are: Host Waits for Another Call, Hot Key to Reactivate Host, Host Accepts Calls at Any Time, and Cancel Host Operations.

Caller Subject to Inactivity Time

Entering Yes in this field causes pcANYWHERE/Host to terminate a remote connection if a caller doesn't enter a keyboard command or perform mouse operations for the number of minutes specified in the Inactivity Timeout field located in the Security section of the Host Preferences menu. Leaving this field to its default No setting allows callers to remain on-line without issuing commands for as long as they like.

Log Calls from this Caller

Entering Yes in this field causes pcANYWHERE/Host to create a log file containing a record of all calls to the host system using this password. Information recorded in this log includes the caller's name and the date and time of each call. This option provides a convenient method for keeping track of remote connections. (A single pcANYWHERE log file is used to record connections for callers using any password with a Yes setting in this field of its Caller Information file.)

Print Destination

One powerful feature of pcANYWHERE is its ability to redirect output from the host to a printer connected to the remote system. This allows a worker in the field, for example, to create a sales report using an application program installed on the host system and send the results to his or her printer. Options available for this setting include Discard Printer Output, Print to Host Only, Print to Remote Only, and Print to Host and Remote.

Note: This setting specifies only the initial destination of print output. The remote caller also can redirect printing using the Print Options on the Remote Session Options menu, which I'll discuss in the next chapter.

Terminal Type

We've pretty much ignored this fact until now, but you don't have to rely on pcANYWHERE/Remote to initiate a remote-computing session with a host system running pcANYWHERE/Host. Virtually any type of system, including many terminals used with mainframe computers, can be used to access a personal computer running pcANYWHERE/Host. (Of course, if you're involved in PC-to-PC communications, it only makes sense to run the appropriate pcANYWHERE module on both systems. Doing so provides certain capabilities not available with other types of connections.)

You can use the Terminal Type field to specify the kind of terminal that will be used by callers signing on to the host system with this password. If you're uncertain of what type of equipment a remote caller will be using, set this field to its Ask Caller option.

Time Allowed Online

Use this field to specify, in minutes, the maximum amount of time callers using this password will be able to remain connected to your host system during a given remote session.

Callback Phone Number

As explained in the previous chapter, if you're using pcANYWHERE over a modem connection, you can instruct your host system to call back a remote system rather than connecting it directly to the host computer. (This callback feature lets you verify that a specific remote system is being used to contact your host computer.) You can use this field, therefore, in conjunction with the prefix and suffix fields below it, to specify the number that will be called whenever someone attempts to sign on to your host system using this password. (This call will be placed following the amount of time you specify in the Delay Before Callback Attempted field using the Host Preferences option. The Prefix and Suffix fields are used to enter any special sequences needed to place a call from your host computer—for example, the 9 required on some PBX systems to access an outside line.

Command to Execute

Use this field to enter a command sequence to be executed whenever someone signs on to the host system using the password associated with this Caller Information file. If the Allow Use of <Ctrl><Break> is set to No, you can use commands entered into this field to specify exactly what happens during a remote session, thus giving you complete control over the activities you'll allow callers assigned this password to perform on your host system.

Because you'll be using the TEST CALL password exclusively for your own remote access to the host system, go ahead and make it as flexible as possible by modifying some of the default settings in its Caller Information file. (In fact, you might want to change its password to something slightly more cryptic, because this will ultimately become your "control" file, so to speak.)

Specifying caller information

Begin by making it possible for you to modify how the host system is set up after you sign off. To allow modification of the host mode from a remote connection, use the arrow keys to highlight Allow Caller to Change Host Mode and press Enter. Then use the down-arrow key to highlight Yes and press Enter.

Next, set pcANYWHERE/Host to log your maintenance calls, using the following procedure: Use the down-arrow key to highlight Log Calls from this User and press Enter. Use the down-arrow key to highlight Yes and press Enter.

Finally, let's set things up so that, by default, application output will be discarded—that is, not printed at the host or remote system. To have it printed at the remote system, because the Print Destination option is already highlighted, press Enter. Use the up-arrow key to highlight Discard Printer Output and press Enter.

Tip: Remember, you can override this setting later using the Print Options available in the Remote Session menu, which you access from the remote system.

Storing a Caller Information file

Now that your personal access file is configured properly, you can return to the pcANYWHERE/Host Main Menu. To save the current Caller Information settings, press Esc. This will return you to the Caller Names listing. To return to the pcANYWHERE/Host Main Menu, press Esc.

At this point, your display should contain the prompt message shown in Fig. 6-4. Don't worry. You haven't committed some egregious error. To be honest, I pulled somewhat of a fast one on you in the previous chapter—with just cause, as I hope you'll soon agree.

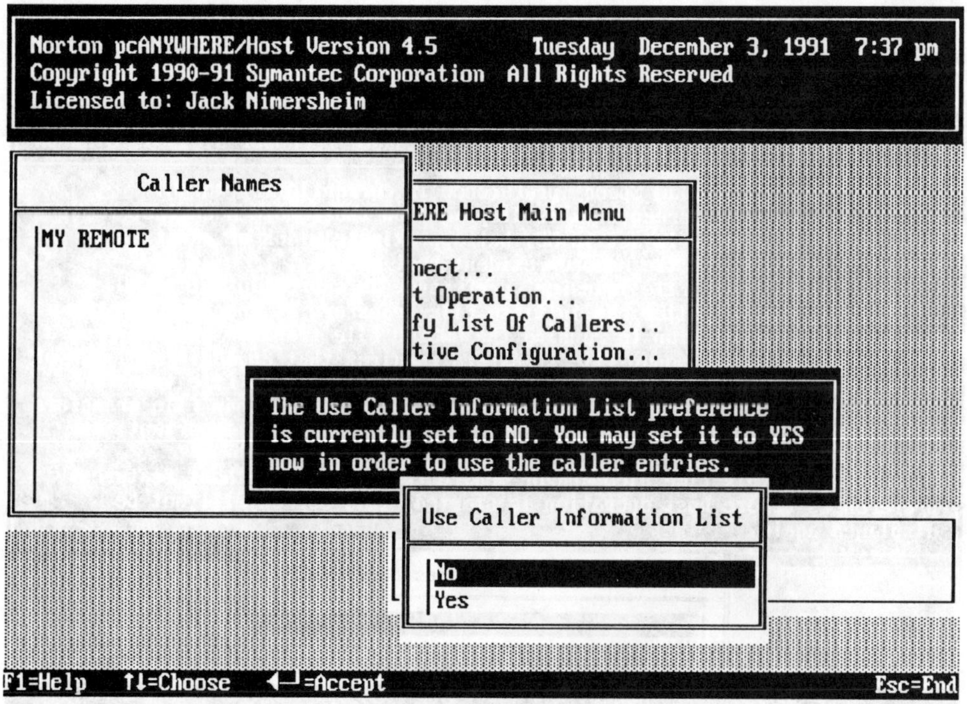

6-4 pcANYWHERE monitors the host environment to ensure against potential conflicts in your program settings.

While you were configuring your default host environment, I didn't have you change the Use Caller Information List field in the Host Preferences section from its default No setting. I did this because I want to show you, now, how pcANYWHERE keeps an eye on what you're doing and, when appropriate, offers suggestions regarding any potentially conflicting settings it detects. In this case, for instance, pcANYWHERE realizes that you just created a Caller Information file, but that you have your host system configured to ignore it. This is just one example of several safeguards pcANYWHERE provides to keep your remote-computing operations running smoothly.

To reconfigure your host system to use the caller information list, use the down-arrow key to highlight Yes and press Enter. After modifying your default configuration to use the caller information list, pcANYWHERE/Host will return you to its Main Menu (see Fig. 6-1).

Beginning host operations

Now that you've created a password file to work with, let's take a look at how you actually prepare pcANYWHERE/Host to accept calls from a remote system. To set your system up to accept a remote call, use the up arrow key to highlight Begin Host Operations and press Enter.

Once again, pcANYWHERE attempts to ensure that your host environment is operating as efficiently as possible by displaying the message screen in Fig. 6-5.

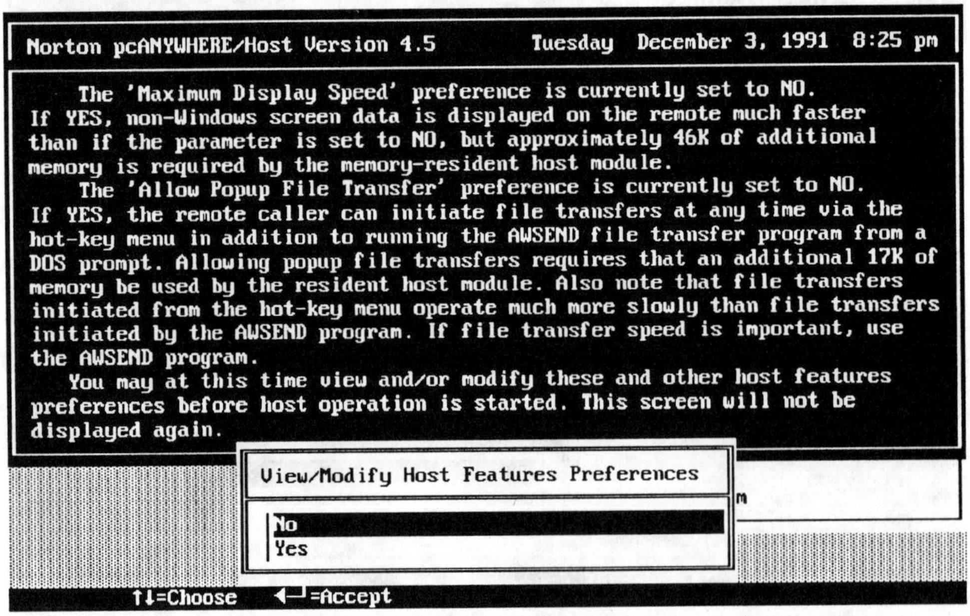

6-5 You can modify pcANYWHERE/Host for improved Windows performance and pop-up file transfers.

This message, which will be displayed only the first time you select the Begin Host Operations option, asks if you want to modify the current Features Preferences settings to speed up remote Windows applications and allow more flexibility in your file-transfer operations. (It also points out approximately how much memory you'll sacrifice to enable these two options.)

For the following exercise, I'm going to assume that you'll be running Windows applications remotely and, furthermore, that you'll want access to pcANYWHERE's pop-up file-transfer utility. Note: If this isn't the case, you can press Enter now to accept No and advance to the Begin Host Operation menu.

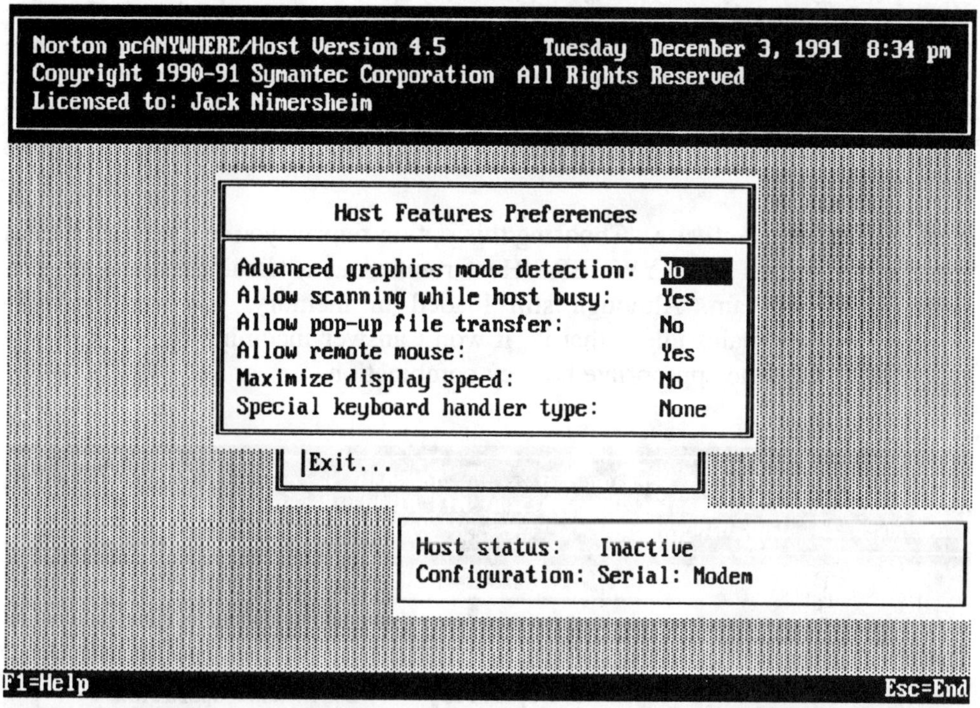

6-6 The Host Features Preferences option box.

To modify the Maximum Display Speed and Allow Popup File Transfer settings, use the down arrow key to highlight Yes and press Enter. Selecting Yes displays the Host Features Preferences option box, shown in Fig. 6-6. This is the same option box you used in the previous chapter to configure pcANYWHERE/Host. To modify your host environment for faster Windows operations and pop-up file transfers:

1. Use the down arrow key to highlight Allow Pop-Up File Transfer and press Enter.
2. Use the down arrow key to highlight Yes and press Enter.

3. Use the down arrow key to highlight Maximize Display Speed and press Enter.
4. Use the down arrow key to highlight Yes and press Enter.
5. Press Esc to exit the Host Features Preferences dialog box.
6. At the Save Changes prompt, press Enter to specify Yes.

After modifying your default configuration to match the new settings, pcANYWHERE/Host will advance you to the Begin Host Operations menu, shown in Fig. 6-7. This menu lists the various options available to you when you decide to prepare your host system to accept calls from a remote computer. They include:

Wait for a Connection Choosing this option dedicates your host system exclusively to pcANYWHERE operations. After initializing your modem, pcANYWHERE/Host displays a message screen indicating it's awaiting incoming calls. Once this message appears, you won't be able to use your computer for any other work until you either cancel host operations or a remote caller attempts to contact the host system.

Exit, Hot Key to Activate Choosing this option returns you to the DOS system prompt, but leaves pcANYWHERE/Host running as a terminate-and-stay-resident (TSR) program. Although still loaded in memory, however, pcANYWHERE/Host remains idle— that is, it won't answer incoming calls—until you reactivate it with the appropriate hot-key combination.

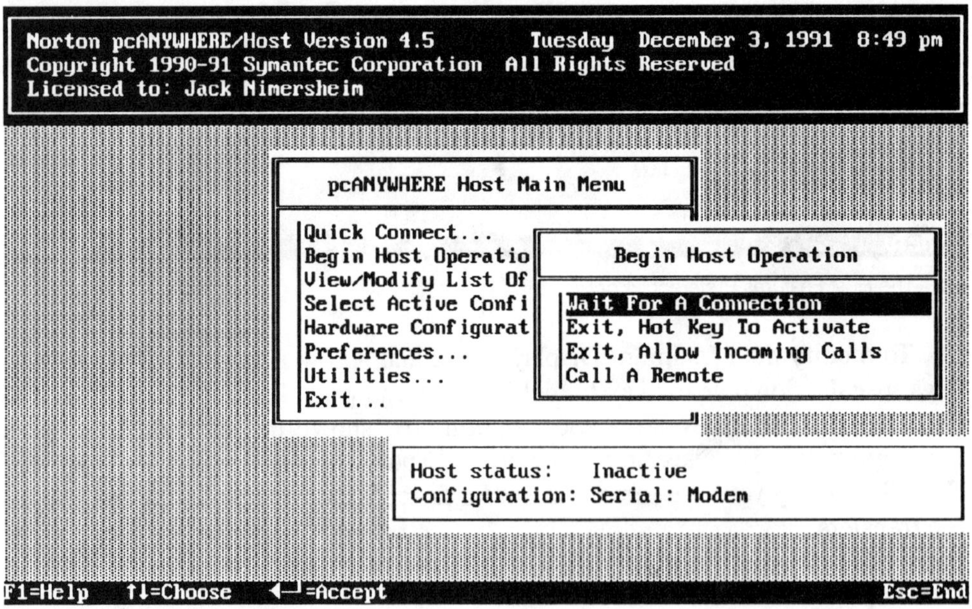

6-7 The Begin Host Operation menu.

Exit, Allow Incoming Calls Choosing this option also sets up pcANYWHERE/Host as a TSR and returns you to the DOS system prompt. In this case, however, pcANYWHERE/Host remains active in the background, ready to automatically answer incoming calls from a remote system.

Call a Remote Although most remote-computing sessions originate on the remote side of the connection, it's also possible to use pcANYWHERE/Host to call another computer. Choosing this option lets you specify the phone number of a remote system you want to call, using the communication parameters specified in the currently active hardware configuration.

Note: At certain times, the Begin Host Operations menu will contain two additional options:

Voice-first Connection When available, you can use this option to begin a remote session between two computers over a line you've already been using for voice communications. This avoids the inconvenience of having to disconnect the line and then tell pcANYWHERE to place a second call.

Cancel and Remove from Memory This option appears whenever the pcANYWHERE/Host program is already active—that is, loaded into memory—and you display the Host menu. Selecting this option causes pcANYWHERE/Host to attempt to remove itself from memory and return you to the DOS system prompt. After selecting this option, the only way to initiate a remote session is to reload pcANYWHERE and reactivate host operations. (Under certain conditions, DOS won't allow pcANYWHERE/Host to be removed from memory. Should this happen, although portions of the program remain in RAM, you'll still need to rerun pcANYWHERE/Host before you can establish a remote connection.)

Summary

What you're going to do now might strike you as being a little odd. Basically, you're going to leave everything just as it is on the host system and move over to the remote side of your pcANYWHERE environment. (Remember, I reminded you earlier in this chapter that every pcANYWHERE session involves two computers.) You'll return to the host system in chapter 8, after a short detour to discuss and fine-tune your pcANYWHERE/Remote environment.

7
CHAPTER

Using pcANYWHERE/ Remote

In this chapter you're going to put the finishing touches on your pcANYWHERE/ Remote environment, the final step in preparing to establish an on-line connection between a host and remote system. Primarily, this involves creating an entry in pcANYWHERE/Remote's list of hosts. This host entry, similar to the caller list entry you created with pcANYWHERE/Host in the previous chapter, will contain the operating parameters you want to use when calling a specific host system. So let's head over to your remote computer and get started. To start pcANYWHERE/ Remote, type AWREMOTE and press Enter.

Note: As before, this step assumes that you're either in the AWR directory created in chapter 4, or that you've added this directory to your DOS search path. If not, you'll need to type CD \AWR and press Enter to make the directory containing your pcANYWHERE/Remote files the active directory before the above command will work.

The AWREMOTE command will load your pcANYWHERE/Remote software and display the Remote Main Menu, shown in Fig. 7-1. You've seen this menu previously, in chapter 5, when you used the Hardware Configurations and Preferences options to establish your default remote settings.

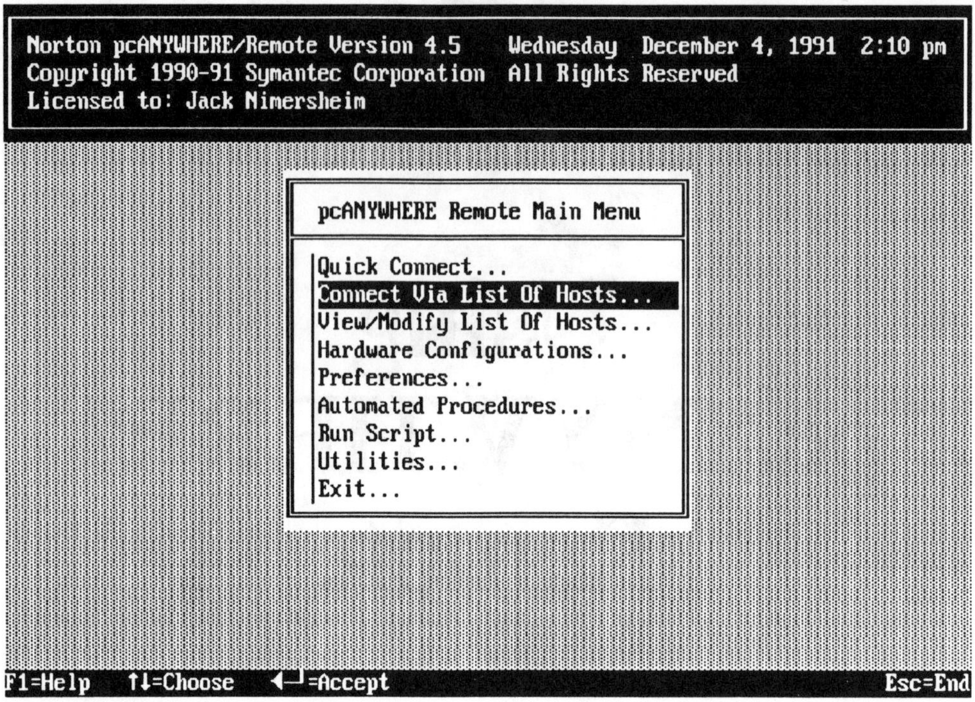

7-1 The pcANYWHERE/Remote Main Menu.

Working with the host list

Most people I know keep track of their planned activities using some kind of appointment book or calendar. Whenever they receive an invitation to a party or some other important event, it gets marked down in this book or calendar to ensure they don't forget it. In keeping with the party metaphor I've been using to represent the host/remote relationship, pcANYWHERE/Remote's list of hosts resembles this familiar organizational tool, in that it allows you to record and maintain information regarding the various host systems for which you have official invitations to visit.

Keeping this metaphor in mind, let's go ahead and create a host entry for the host system you configured in the previous chapter.

Adding entries to the host list

Use the View/Modify List of Hosts option on the pcANYWHERE/Remote Main Menu to access the list of host systems you plan to contact from your remote computer. To access the pcANYWHERE/Remote host list, type V and press Enter, or use the arrow keys to highlight View/Modify List of Hosts and press Enter.

Selecting this option causes pcANYWHERE/Remote to display the names of any individuals currently entered in your caller list. Immediately following the installation of pcANYWHERE/Remote, this list contains two entries, as illustrated in Fig. 7-2: pcANYWHERE Host and Non-pcANYWHERE Host.

The only initial difference between these two entries is their terminal emulation settings—that is, how each handles keyboard entry and manages interaction between your PC's display and that of the remote system during an on-line session. The latter emulates DEC VT-100, a terminal type commonly used on mainframe and minicomputers, while the former uses pcANYWHERE's proprietary terminal settings to improve the efficiency of a remote connection between two computers that are both running the pcANYWHERE software.

Notice that, as was true with the pcANYWHERE/Host list of callers, the various options available to you when working with the host list are listed across the bottom row of Fig. 7-2. You'll use the Add option to create a new host entry.

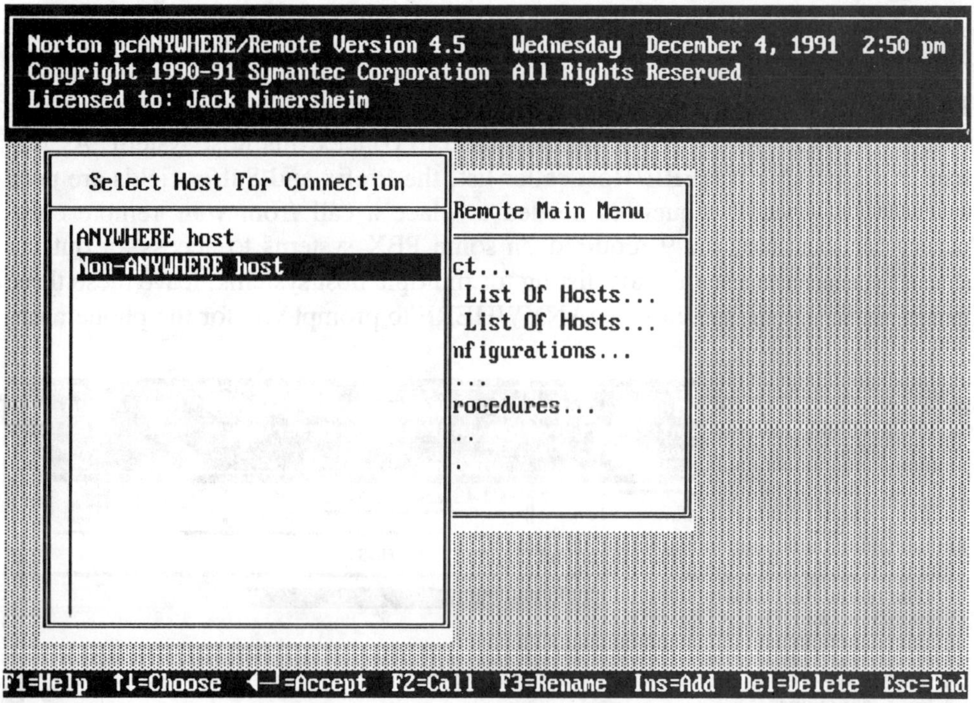

7-2 Adding entries to pcANYWHERE/Remote's list of hosts simplifies the process of contacting host systems.

To add a new entry to your host list, press Ins to specify a new entry. Note: Because adding new entries to the host list represents a modification to your remote environment, pcANYWHERE won't allow you to continue past this point without entering the master password you assigned to your remote system in chapter 5. Go ahead and do this now, then press Enter.

Host Information files

Selecting the Add option displays a prompt box with which to name your new host entry. As pcANYWHERE/Host did on your host system, pcANYWHERE/Remote uses this information to maintain a log of your on-line activities on the remote computer. As was true for your remote caller list, therefore, you should pick names that allow you to easily identify each host system you call. To assign this entry a host name, type MY HOST and press Enter.

After you assign a name to your new host entry, pcANYWHERE will advance you to the dialog box shown in Fig. 7-3. Use this dialog box to specify the procedures used to contact and communicate with the host system associated with this listing whenever you select it from your host list. Items specified in this dialog box include:

Configuration Name Use this field to specify the hardware configuration containing the communications parameters you want in effect whenever you contact this host.

Phone Number Use this field, in conjunction with the Prefix and Suffix fields below it, to specify the phone number used to contact this host system. As was true in the pcANYWHERE/Host caller list, the Prefix and Suffix fields are used to enter any special sequences needed to place a call from your remote computer—for example, the 9 required on some PBX systems to access an outside line. If you plan to use this host listing for multiple host systems, leave these three fields blank. Doing so causes pcANYWHERE to prompt you for the phone num-

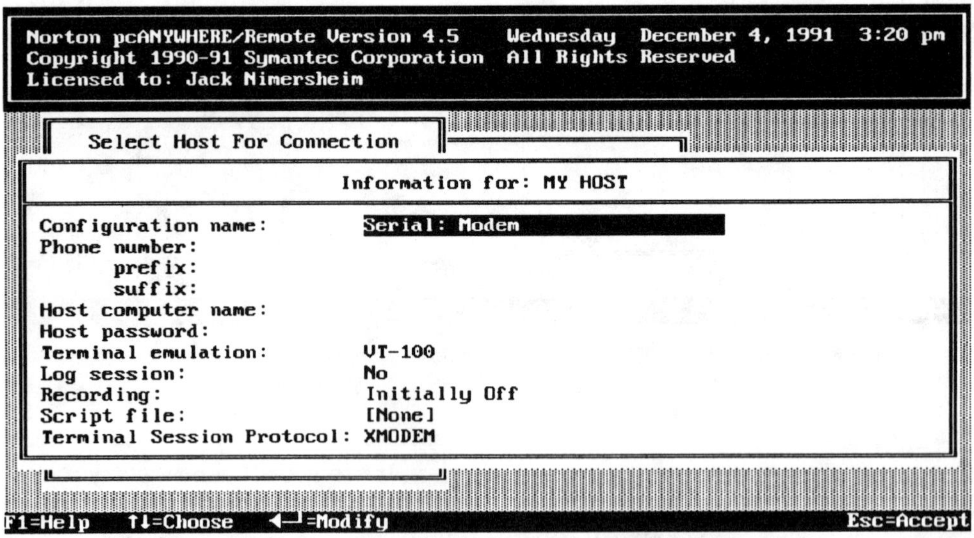

7-3 The Information dialog box displays information about the current host settings.

ber you want to call, whenever you try to contact another computer using this host listing.

Host Computer Name Use this field to specify the name assigned to the host system you'll be calling. If you plan to use this host listing for multiple host systems, leave this field blank. Doing so causes pcANYWHERE to display a list of known hosts whenever you try to contact another computer using this host listing.

Host Password If this host listing is being used to call a single remote system, enter the password you've been assigned for that system here. pcANYWHERE will automatically transmit this password during sign-on. If you plan to use this host listing for multiple host systems, leave the password field blank. If pcANYWHERE/Remote does not automatically transmit a password, the host system will prompt you to enter one during sign-on.

Terminal Emulation As mentioned earlier, this setting determines how pcANYWHERE handles keyboard entry and manages interaction between your PC's display and that of the remote system during an on-line session. You should always set this to pcANYWHERE when calling a system on which pcANYWHERE/Host is running. For non-pcANYWHERE systems, you'll need to pick the appropriate terminal emulation for the type of system it is.

Log Session Entering Yes in this field causes pcANYWHERE/Remote to keep a record of the date, time and duration of any call made to this host. (The View Log Files option in the pcANYWHERE/Remote Utilities menu allows you to request various reports containing information on calls placed from this remote computer.)

Recording Have you ever used a cassette recorder to record an important meeting? pcANYWHERE allows you to do much the same thing with your on-line sessions by saving all screen activity on the remote system to a file. A special Play Back option in the pcANYWHERE/Remote Utilities menu can be used to review the contents of a session recording. (The value in this field specifies only the initial status of the Recording option each time you connect with this host. Alternately, you can manually turn recording on or off at any time during an on-line session.)

Script File I haven't discussed scripts yet. They're one of the truly impressive features of pcANYWHERE that I'll be covering extensively in Part 3 of this book, *Beyond the basics*. Stated simply, scripts written using pcANYWHERE's procedures language can be used to automate your on-line activities—that is, cause them to be executed without any intervention from you. Use the Script File field to identify an optional script you want run each time this host listing is used to contact another computer.

Terminal Session Protocol This field identifies the default file-transfer protocol that will be used when terminal emulation is in effect—that is, when a setting other than pcANYWHERE is specified in the Terminal Emulation field. The contents of this field are ignored during remote sessions in which both computers are running pcANYWHERE.

Defining host settings

Now that you know how pcANYWHERE uses each field in the host list Information dialog box, let's go ahead and set up this new listing to reflect the settings you've specified for your host system in the previous two chapters.

To create an information file for your host system, use the down arrow key to highlight the Phone Number field. Then type in the phone number for your host system and press Enter. (Don't forget to fill in the Prefix and Suffix fields, if necessary.) Then:

1. Use the down arrow key to highlight the Host Computer Name field.
2. Type MYHOST and press Enter.
3. Type TEST CALL and press Enter.
4. Press Enter to display the Terminal Emulation option list.
5. Use the up arrow key to highlight pcANYWHERE and press Enter.
6. Press Enter to display the Log Session option list.
7. Use the down arrow key to highlight Yes and press Enter.
8. Press Esc to save the new settings and exit the host list Information dialog box.

This will return you to a host list similar to the one shown in Fig. 7-2. Notice that this list now includes your MY HOST entry. Also notice that, instead of simply displaying your host list, pcANYWHERE is now asking you to select a host for the current connection, as illustrated in Fig. 7-4. That's because, after saving the new settings specified in the above exercises as the new MY HOST listing, pcANYWHERE/Remote figures that you're ready to go on-line and automatically advances you to the Connect via List of Host option on its Main Menu. (Go ahead and press Esc to return to the Main Menu and check this out if you don't believe me.) Because pcANYWHERE/Remote is nice enough to lead you here, why not go ahead and see what happens when you select MY HOST from the host list.

To use the MY HOST listing to establish a remote connection, use the arrow keys to highlight MY HOST and press Enter. Selecting a host listing advances you to the pcANYWHERE/Remote Connect Options menu, shown in Fig. 7-5. This menu lists the various options available to you when you decide to use your remote computer to initiate a remote session.

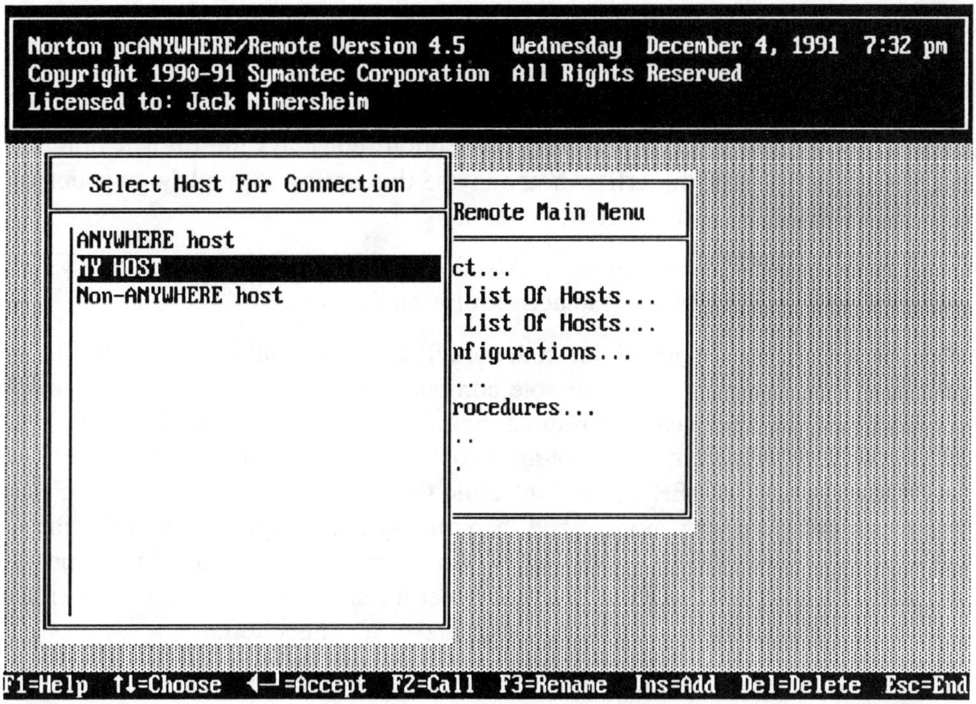

7-4 Selecting a name from the list of hosts applies the settings in its information file to the current session.

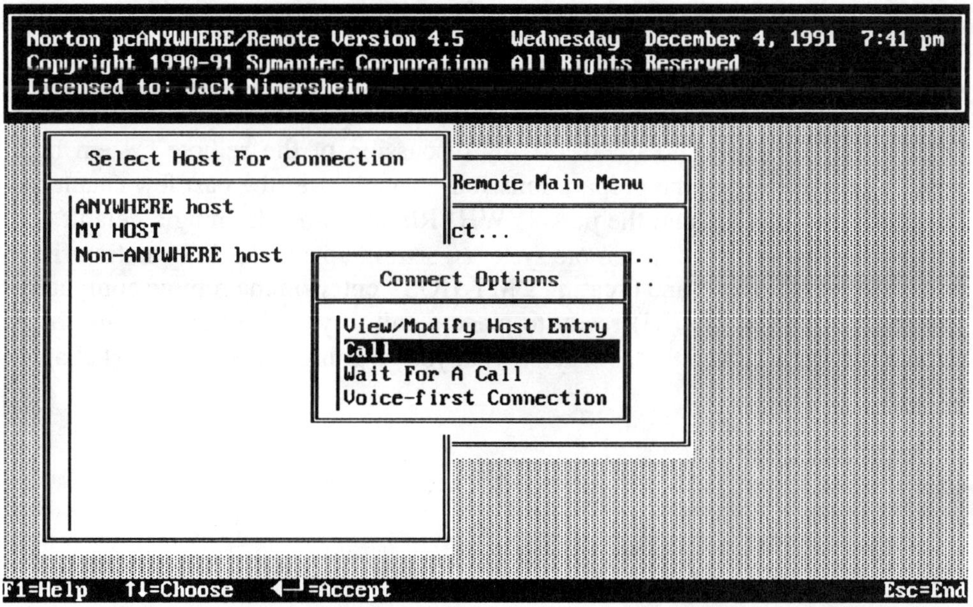

7-5 The pcANYWHERE/Remote Connect Options menu.

pcANYWHERE/Remote Connect Options

Choices available on the pcANYWHERE/Remote Connect Options menu include:

View/Modify Host Entry Selecting this option redisplays the Information dialog box you worked with earlier. You can use this option to modify your default MY HOST settings for the current session.

Call Selecting this option causes pcANYWHERE/Remote to call a host system using the settings specified in the active information file.

Wait for a Call As I pointed out in the previous chapter, although you'll usually contact a host system from the remote computer, it's possible to begin a remote session by having a host call the remote. Selecting the Wait for a Call options prepares your remote system to accept an incoming call from another computer. As was true in pcANYWHERE/Host, selecting this option causes pcANYWHERE/Remote to initialize your modem and then display a message screen indicating to the remote computer that it's awaiting incoming calls. Once this message appears, you won't be able to use your remote computer for any other work until you either cancel remote operations or a host system makes a connection.

Voice-First Connection This option corresponds to the same feature in pcANYWHERE/Host, in that it lets you begin an on-line session between two computers with voice communications and then switch to actual remote computing. The voice-first option works only when both systems involved in a remote session are running pcANYWHERE.

Summary

Once again, I'm going to break off my discussion of the remote system here. Why? Because after all the preparations you've made over the past few chapters—installing your modem and the pcANYWHERE software, defining a default configuration for your host and remote systems, setting up a caller information file in pcANYWHERE/Host, and creating a MY HOST entry on the remote computer—you're finally (Yes, *finally*!) ready to place a call to your host system and begin remote computing. And that is exactly what you're going to do in the next chapter.

8 CHAPTER

Placing that first call

You're about two pages away from using pcANYWHERE to establish a remote connection between two computers. Before taking this important step, however, I thought it might be a good idea to review—very briefly, I assure you—what you've accomplished up to this point. Doing so will make sure that I've avoided the proverbial "forest for the trees" syndrome, where you become so involved with the minutiae of a given project that you lose sight of how the individual pieces fit together within the larger picture. I just want to make certain that I haven't fallen into this all-too-common trap in the process of explaining how to install and configure pcANYWHERE.

Where you stand

You might not realize it but, over the past few chapters, you've built the foundation of an extremely powerful and flexible telecommunications environment. How powerful? How flexible? Let's see if I can answer these two questions.

At the most basic level, pcANYWHERE is now set up to let you take over and run one PC (your host system) using a second PC (the remote computer) in a different location. No surprises here. After all, remote computing has always been the primary function of pcANYWHERE. What you might not realize, however, is that, as I pointed out at the beginning of this book, the same hardware and software that allows you to use pcANYWHERE for remote computing opens an electronic doorway to a wealth of other on-line opportunities and services, as well.

You'll step through that doorway in later chapters, as I explain how pcANYWHERE can be used to contact commercial information services such as Compu-

Serve, Delphi, and BIX, or to send and receive correspondence through an e-mail service like MCI Mail, or to go on an electronic shopping spree without setting foot outside your home. These are but a few of the activities that fall under *The Complete Communications Guide* portion of this book's title. They are also, as I stated earlier, the types of potential uses for pcANYWHERE that, understandably, receive only cursory coverage in the program's otherwise excellent documentation.

After seven chapters of unavoidable delays, you now stand on the threshold of that electronic doorway. So let's go ahead and start opening it. You'll begin, logically enough, with a demonstration of the remote-computing capabilities upon which pcANYWHERE has built its current reputation.

Calling a host system

Let's quickly recap where you left things at the end of the previous two chapters. Specifically, the Begin Host Operation option menu, shown in Fig. 8-1, was displayed on your host system. Over on the remote side of the fence, you should see pcANYWHERE/Remote's Connect Options menu, shown in Fig. 8-2. (If this isn't the case on either system, go ahead and make it so, as they say on the bridge of the Starship Enterprise. I'll wait.)

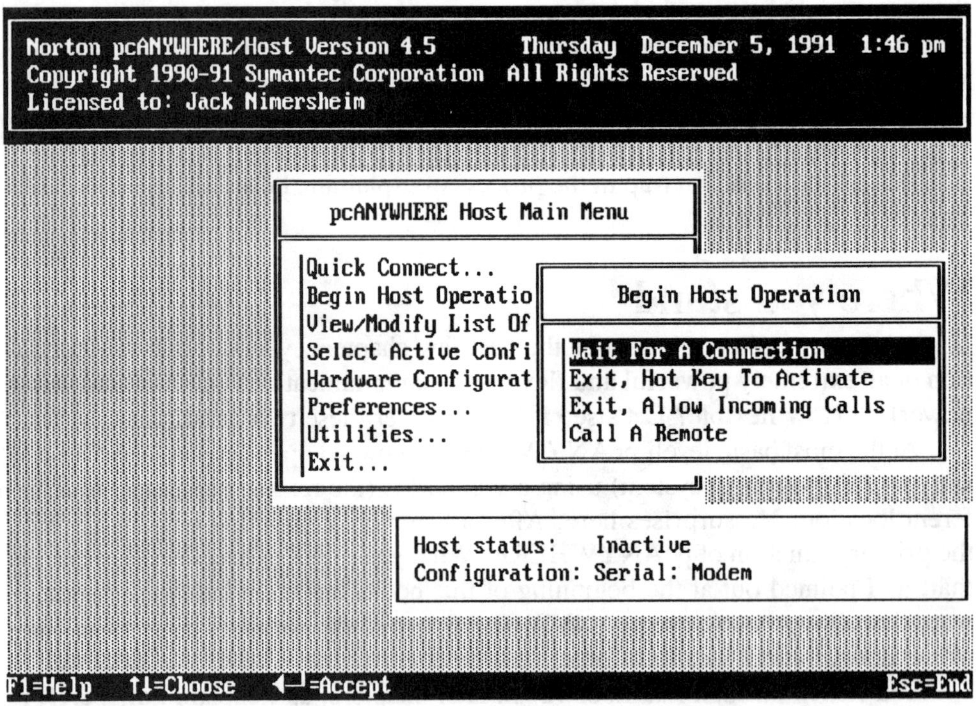

8-1 pcANYWHERE/Host.

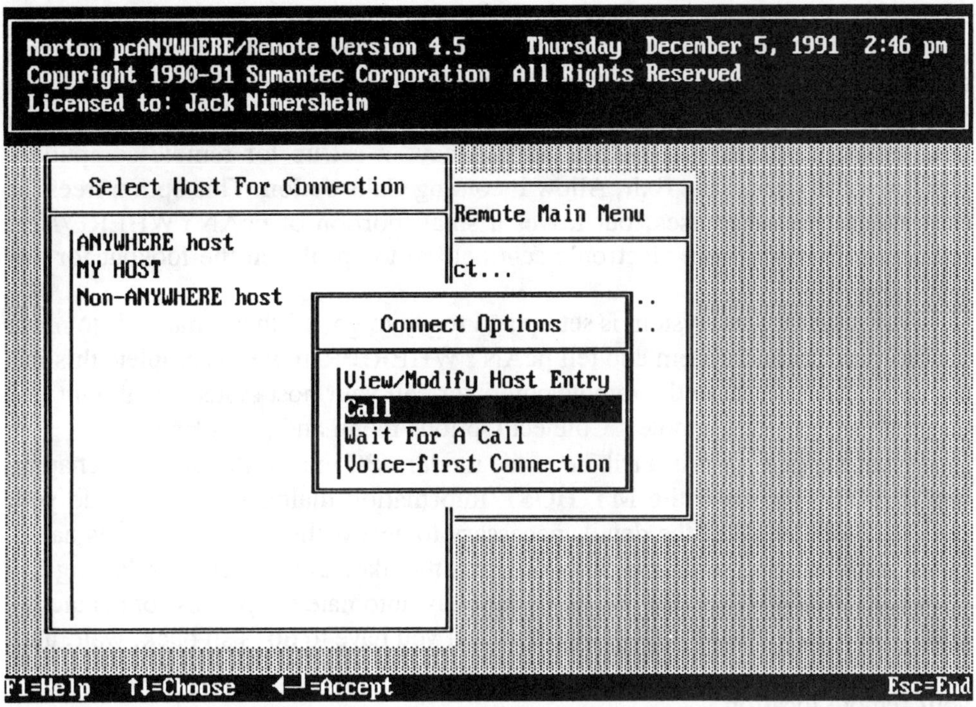

8-2 pcANYWHERE/Remote.

The first step in preparing for an actual remote-computing session involves setting up your host system to automatically answer incoming calls and, ultimately, turn control over to the remote computer.

To prepare the host system to accept incoming calls, highlight Wait for a Connection and press Enter. After a few seconds—during which pcANYWHERE/Host, in essence, takes over your host system and initializes its modem to the settings specified in the active hardware configuration—you'll see the message box shown in Fig. 8-3. As this message indicates, the host system is now set up for remote computing, patiently awaiting an incoming call.

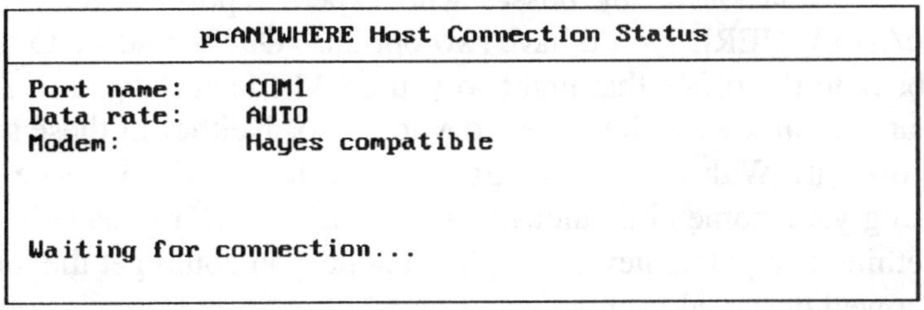

8-3 Your host system is now awaiting an incoming call.

Note: As mentioned in chapter 6, the Wait for a Connection option causes pcANYWHERE to completely take over the host system, dedicating it exclusively to remote computing. In other words, your host system will be unavailable for other activities, unless and until a remote connection is established. If you ever need to work on the host system but still have it set up for remote computing, you'll want to choose the Exit, Allow Incoming Calls option. This option frees the host system for other uses, but leaves a small portion of pcANYWHERE/Host resident in memory—an electronic sentinel, so to speak, on the lookout for any incoming calls.

Now that the host system is set up and ready to go, all that remains is to move over to your remote system and tell pcANYWHERE/Remote to complete this initial connection. To have the remote computer call your host system, highlight Call on pcANYWHERE/Remote's Connect Options menu and press Enter.

What happens next is a tribute to the modern PC era. In the previous chapter, you'll recall, you used the MY HOST Information dialog box to provide your remote computer all of the details necessary to access the host system—its name, phone number, your assigned password, and the like. Consequently, selecting Call allows pcANYWHERE/Remote to completely automate the process of contacting and signing on to the MY HOST system. All you have to do is sit back, wait, and, a few moments later, you'll be able to assume total control of the host system from your remote location.

After all your waiting, this process seems somewhat anticlimactic, doesn't it? With the appropriate preparation, however, using pcANYWHERE to establish a remote connection between two computers really is this simple. However, you've only begun to examine what you can do once such a connection exists. For starters, you can run virtually any program installed on the host system from the remote location. This, perhaps more than any other pcANYWHERE feature, is what makes remote computing so attractive.

What if, for example, your electronic companion on a business trip is a laptop computer that lacks the system resources required to run a truly high-end application you rely on to complete critical portions of your work? A quick call to an adequately equipped host system would resolve this problem.

Or what if, after arriving home from the office one evening, you realize that you forgot to print a report the boss wants on his desk "first thing tomorrow morning?" (Do you know any bosses who expect reports at any other time?) Without pcANYWHERE, you'd have two options. One would be to turn around and head back to the office that night so you could finish things up. Alternately, you could arrive early enough the next day to do so. Neither of these sounds very attractive, does it? With pcANYWHERE, you'd have a third, more convenient option. Using your home computer to call the office PC and have it work overtime, something computers never complain about, you could get the job done and still enjoy a well deserved evening off.

Put your mind to it, and I'll bet you can come up with at least a dozen more "what if" scenarios where pcANYWHERE's ability to run application programs over a remote connection could prove invaluable.

The Session Options menu

Adding to pcANYWHERE's value is the fact that, while you're using it, you actually have access to the resources of two computers. On the surface, it might appear as if the remote computer is dedicated solely to controlling your host system. This isn't the case at all, however, as a quick keystroke will demonstrate. To see what I mean, press Ctrl−Right Shift.

Note: To avoid potential conflicts with application programs running on your host system, pcANYWHERE uses multi-key commands—that is, pressing two or more keys at the same time—to initiate certain operations during a remote-computing session. The standard notation for indicating such commands, which I'll use throughout this book, is to list all of the keys that must be pressed, separating each with a dash. The previous command, for example, is telling you to hold down the Ctrl key while pressing the Shift key on the right-hand side of your keyboard.

Two things should have happened when you entered this command: 1) The host system displays the message, Remote user is busy. Please wait . . ., and 2) The remote computer displays the pcANYWHERE Session Options menu, shown in Fig. 8-4.

8-4 The pcANYWHERE/Remote Session Options menu.

The meaning of the message displayed on the host system is fairly obvious—that the person on the other end of the connection has temporarily suspended pcANYWHERE to work exclusively on the remote system. The new display on the remote system, the Session Options menu, lists what types of operations the remote user can choose to perform while on-line during a pcANYWHERE session.

Note: Figure 8-4 contains a complete listing of pcANYWHERE's Session options. Not all of these options will show up or, alternately, be functional in the Session Options menu for every caller. pcANYWHERE determines which Session options a given remote user can access, based on the host preferences associated with his assigned password.

You'll experiment with the various Session options in the next chapter. For now, let me briefly describe how each is used:

Chat with Host User This option allows the two computers involved in a remote session to communicate directly with one another. Selecting this option causes two display windows to appear on both computers—one containing typed comments from the remote user and a second for a person at the host system. This option provides a convenient method for the host and remote users to coordinate activities during a remote session.

End Session Use this option to terminate the current connection to the host system. Before dropping the connection, pcANYWHERE presents you with a second list of options, from which you can choose how the host system should be configured at the end of the current remote session. Which options you're presented with depends on what privileges are associated with your password.

File Transfer Use this option to exchange files between the host and remote systems. Transferring files from the Session Options menu takes longer than using pcANYWHERE's AWSEND program, introduced later in this chapter. Because this option operates in the background, however, it allows you to continue working on the host system while the file transfer occurs. (The File Transfer option is functional only if the host preferences are set to allow background file transfers.)

Go to DOS This option temporarily suspends remote access and allows the person sitting at the remote system to work directly from his or her DOS system prompt. Commands entered after selecting this option are executed on the remote, rather than the host, computer.

Mouse If the host system is set up to support mouse commands from the remote computer, you can use this option to enable or disable the remote mouse and adjust the speed of its operations.

Print Options Use this option to specify a destination for print operations initiated during a remote session. Output can be sent to a disk file or a printer connected to either the host or remote system.

Reboot Host Computer This option, which can be useful if an application program malfunctions on the host system, lets you restart the host computer from a remote location. The host system can be configured to support or disable this option for specific callers. (The appropriate command to restart the pcANYWHERE/Host software must be included in the host computer's AUTOEXEC .BAT file for a caller to reestablish a remote connection following a reboot.)

Recording On As mentioned earlier, pcANYWHERE/Remote lets you record your remote sessions to a disk file for subsequent review. This option allows you to toggle session recording on and off. Information in the host's Caller Information file determines whether session recording automatically begins when a caller establishes a remote connection.

Refresh Screen Selecting this option directs the host system to resend its current screen contents to the remote computer. Although it happens only rarely, a screen image from the host can become garbled—that is, include invalid characters—during transmission to the remote session. Should this occur, selecting this option will correct the problem.

Run Script As I mentioned earlier, pcANYWHERE's script language allows you to automate your on-line activities. (Scripts are one of the topics I'll cover in Part 3 of this book, *Beyond the basics*.) You use the Run Script option to execute script files stored on the host system from a remote location.

Save Screen This option lets you save the current contents of the remote display to a disk file for future reference. This differs from session recording in that it represents a "snapshot" of a specific display rather than an ongoing log of remote activities. If you plan to capture more than one display during a remote session, additional screen shots can be appended to an already-existing file.

Special Options Selecting this option displays a dialog box containing several settings that influence how the host and remote computers interact with one another. The primary reason for changing these settings is to compensate for differences between the hardware installed on the two systems involved in a remote session. For example, it might be necessary to modify certain screen attributes when a remote computer with a monochrome display is used in conjunction with application programs set up to run in color on the host.

Switch to Voice This option temporarily suspends a remote session and allows the host and remote users to talk with one another, using voice communications—that's computerese for talking on the phone, folks. For this option to work properly, the two telephones must be connected to extension plugs on each system's modem. (It goes without saying that the Switch to Voice option is available only with pcANYWHERE running over a modem connection.)

Note: The host system also includes a Session Options menu. It, too, can be invoked by a user at the host location using the appropriate hot-key combination during a remote session. The host Session options, however, are limited to the following four items:

- Chat with Remote User
- Print Options
- End Session
- Special Options

Each of these four selections performs the same function on the host system as its counterpart on the remote computer's Session Options menu.

As I said at the beginning of this section, you'll be looking at individual Session Options in the next chapter. For the remainder of this chapter, let's go ahead and perform some basic operations on the host system, using pcANYWHERE/Remote as you normally would, once a remote connection has been established.

Remote possibilities

To reprise a theme introduced way back at the beginning of chapter 1, you are now, for all intents and purposes, "in two places at the same time." Although you're sitting at the remote computer, any commands you enter are actually executed on a host system, which could be a few feet or several hundred miles away. To get a better idea of what I mean, try the following experiment:

1. Type CLS and press Enter to clear the screen.
2. Type CD \AWH and press Enter to make AWH the active directory.
3. Type DIR /W and press Enter.

Your screen should now resemble Fig. 8-5. This might look like a typical DOS directory listing, not that different from what you'd see whenever you enter a DIR command on your remote computer. Look closely, however, and you'll realize that the files listed in Fig. 8-5 are the programs that pcANYWHERE's installation utility installed on your host system back in chapter 4. AWHOST.EXE, for example, is the program you ran to set up your host system for remote computing.

Now, type TYPE READ.ME|MORE and press Enter. After entering this command, the display on your remote system will resemble Fig. 8-6, clearly identified as the release notes for Norton pcANYWHERE/Host, version 4.5. If this doesn't dispel any doubts as to whether the commands entered on your remote computer are actually being executed on the host system, I don't know what will.

Note: The actual contents of your READ.ME file might differ slightly from those shown here. Release notes will change as a company like Symantec constantly modifies and refines its products. What's important to note is the file's identification, located at the top of Fig. 8-6.

To end this experiment, press Ctrl−C. This terminates the TYPE operation and returns you to the DOS system prompt. (In chapter 6, you might recall, you configured the Caller Information file on your host system for this session to allow

```
C:\>CD \AWH

C:\AWH>DIR /W

 Volume in drive C is ZENITH 286
 Volume Serial Number is 16AD-4C0B
 Directory of C:\AWH

[.]             [..]            AW.CI1          AW.HC1          AW.HW1
AW.TRM          AWDVR_0.OVL     AWDVR_1.OVL     AWDVR_C.OVL     AWDVR_G.OVL
AWEGA.DRV       AWH.MSG         AWHOST.EXE      AWHOVL1.EXE     AWKBD.DRV
AWKBDHP.DRV     AWKBDOLI.DRV    AWLOGOFF.EXE    AWMOUSE.DRV     AWSEND.EXE
AWSEND.OVL      AWVCD.386       AWVDDEGA.386    AWVDDVGA.386    AWVGA.DRV
BTRIEVE.EXE     INSTALL.EXE     INSTALL.MSG     READ.ME         AW.HL1
       30 file(s)      990742 bytes
                     25991168 bytes free

C:\AWH>
```

8-5 A directory listing from the host system.

```
                Release Notes for Norton pcANYWHERE/Host, version 4.5

    *************************** WINDOWS **********************************

o   The "pcANYWHERE Installation for Windows" procedure removes a reference
    to a driver named VIPX.386 from the "network=" line in your SYSTEM.INI
    file. This line appears if a Novell network was specified when Windows
    was originally installed. Current versions of vipx do not work and will
    cause pcANYWHERE to lock up. Future versions of VIPX.386 may work
    properly and may be needed by other applications. If this is the case,
    add the reference back into the SYSTEM.INI file.
    The original line in the SYSTEM.INI contained something like:
         network=*vnetbios, vnetware.386, vipx.386
    The reference to vipx.386 is removed by the installation program:
         network=*vnetbios, vnetware.386

                                                                <continued>
    ************************** WINDOWS (continued) **********************

o   When used with Windows, pcANYWHERE 4.5 supports EGA and VGA only.
    If you have installed a Windows Super-VGA driver, Windows will still
    work in Super-VGA mode when executed on the host with pcANYWHERE not
    loaded. Howver, if Windows is run remotely, only the standard VGA
-- More --
```

8-6 DOS commands issued on the remote computer execute on the host system.

a remote user to use either a Ctrl – C or Ctrl – Break key combination to interrupt DOS operations.) Once again, the command you enter at the remote computer's keyboard is executed on the host system.

At this point, I'm faced with an unusual dilemma. Normally, whenever I write a book about a computer program, I can present a series of exercises designed to provide specific examples of the program's major features. pcANYWHERE's primary purpose, however, is to let you run *other programs* over a remote connection. My host system, for example, contains a number of popular PC programs—WordStar, Windows, and PC Tools, to name but three—any of which pcANYWHERE allows me to control from my remote system. Unfortunately, other than DOS (which you've used in the previous two exercises) and pcANYWHERE itself, I have no way of knowing what software is installed on your host system. Consequently, I can't demonstrate how to run them remotely.

Don't misunderstand. There are still a lot of pcANYWHERE features we can explore together. (There better be; I have a couple hundred pages to fill.) When it comes to using pcANYWHERE with the actual applications on your host system, however, you're pretty much on your own.

My recommendation? Take several minutes now and try running some programs already installed on your host system. Pick a few of your favorites and put them through their paces. That's really the only way you'll discover the possibilities inherent in remote computing. When you're done experimenting, head back here and we'll close out this chapter by looking at a feature of pcANYWHERE that everyone can take advantage of, its AWSEND file-transfer utility.

The AWSEND file-transfer utility

One of pcANYWHERE's more convenient features is its ability to transfer files between your host and remote computers. If you think Federal Express's much-touted promise to deliver something from one place to another "absolutely, positively" overnight is impressive, wait 'til you see pcANYWHERE's AWSEND program in action. If the "something" you need to deliver is a program or data file, AWSEND can move it on its way at speeds that put to shame even the fast-talking pitchman who first introduced Fed Ex's now-famous motto to the American lexicon.

The most impressive aspect of AWSEND is not its speed, however; almost any communications program will allow you to transfer files quickly over a modem connection. The true beauty of AWSEND is that it converts this often confusing activity into a simple, menu-driven procedure that even first-time users should have no trouble mastering. But rather than merely talk about AWSEND, why don't I let you see it in action. Then you can judge for yourself how fast and convenient the particular pcANYWHERE feature is to use.

Starting AWSEND

Because I don't know exactly what you did during your initial experiments with remote-computing, I want to make sure that everyone reading this book begins in the same place at the same time, relative to their host system's disk and directory structure, as we begin our examination of AWSEND. The best way to guarantee this is to clear the screen and get back to the AWH directory you created when you installed pcANYWHERE on your host system. To return to the AWH directory:

1. Type CLS and press Enter to clear the screen.
2. Type C: and press Enter to make drive C your active disk drive.
3. Type CD \AWH and press Enter to make AWH (the pcANYWHERE/Host directory) your active directory.

Now that the playing field is level, as they say in political circles, let's go ahead and see how easy it is to use AWSEND, pcANYWHERE's file-transfer utility. To run AWSEND, type AWSEND and press Enter. After a few seconds, your screen will resemble Fig. 8-7. This is the main AWSEND display, which can be used to initiate file transfers between the host and remote computers.

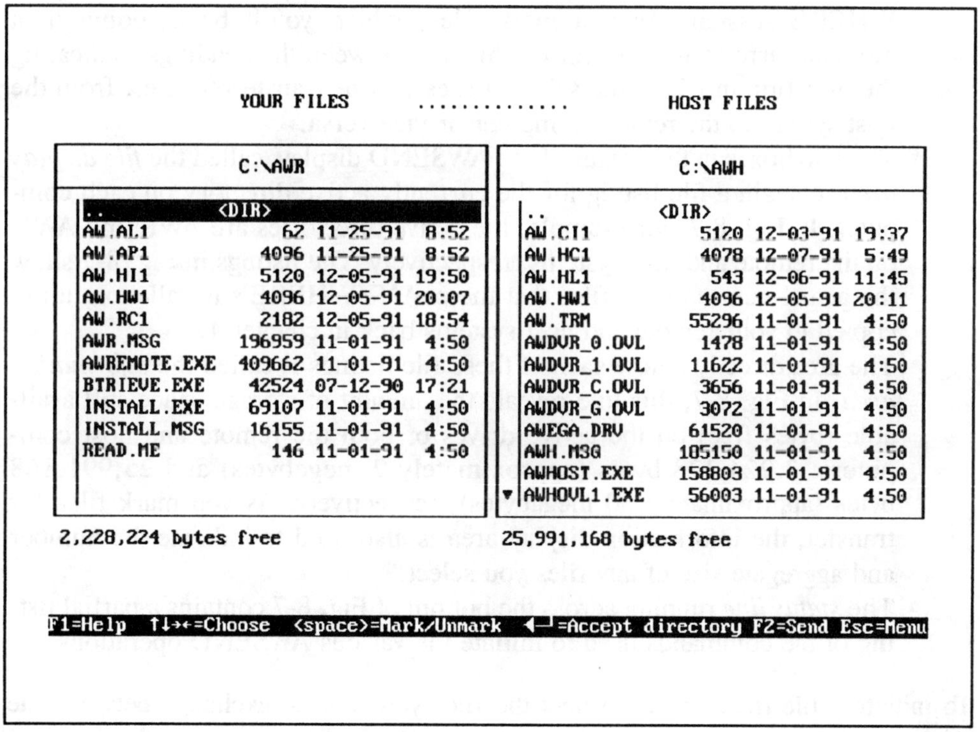

8-7 The AWSEND display.

Note: Running AWHOST from the remote computer causes the host system to display a message box indicating that a file transfer is in progress. If the currently active hardware configuration doesn't disable the host system's keyboard during a remote connection, the host user can cancel AWSEND by pressing Esc.

As is true of most instructions entered at a remote computer, the previous command actually executes a program on the host system—in this case, the AWSEND file-transfer utility. Don't look for AWSEND.EXE on your remote computer's disk; it isn't there. Even though file transfers can be initiated from either the host or remote computer, the AWSEND program itself resides exclusively on the host system. A quick analysis of Fig. 8-7 will not only verify this fact, it will also provide a basic understanding of how you use AWHOST to exchange files between the two computers involved in a pcANYWHERE session.

The AWSEND display

As Fig. 8-7 illustrates, the AWSEND display is divided into four main sections:

- The top line of the AWSEND display contains the headings YOUR FILES and HOST FILES, identifying the two systems involved in a pcANYWHERE session. As you mark files, which you'll begin doing in a moment, arrows will appear on this line between the headings, indicating the direction in which the selected files will be transferred, i.e., from the host system to the remote computer or vice versa.
- The two boxes in the center of the AWSEND display, called the *file display area*, contain a file listing for the currently active directory on each computer. In Fig. 8-7, for example, the active directories are AWR and AWH on the remote and host systems, respectively. The listings in Fig. 8-7 show the actual names of the files that the pcANYWHERE's installation utility copied to your remote and host systems back in chapter 4.
- The area directly below each of these file listings is called the *information area*. In Fig. 8-7, this line reveals the amount of storage space still available (bytes free) on the active drives of both the remote and host computers: 2,224,128 bytes (approximately 2 megabytes) and 25,991,168 bytes (approximately 30 megabytes), respectively. As you mark files for transfer, the information display area is also used to indicate the number and aggregate size of any files you select.
- The *status line* running across the bottom of Fig. 8-7 contains a partial listing of the commands used to initiate the various AWSEND operations.

To initiate a file transfer, first select the files you want to exchange between the host and remote systems. Using the standard DOS COPY command, this involves manually identifying individual files or a group of files to be duplicated.

AWSEND greatly simplifies this process by allowing you to visually mark the files you want to transfer from one computer to another.

The direction of a file exchange, host-to-remote or remote-to-host, depends on which box within the file display listing you use to mark the file or files you want to transfer. Files you mark in the remote box will be copied to the host system. Conversely, marking files in the host box causes the selected files to be copied to the remote system. It's possible to select multiple files for transfer, but you can copy files in only one direction during a given copy operation.

Demonstrating how a file transfer works will be much more effective than merely talking about it. In the following exercise, you'll take several files on the host system and copy them to a new directory on the remote system's hard disk. In the process, you'll see how the AWSEND Options menu and various keyboard shortcuts are used to prepare for and initiate a file transfer.

Selecting a file

Begin by making HOST FILES the active file display listing and then marking a single file in the AWH directory. To select a file on the host system, press the right arrow to move the highlight bar to the HOST FILES file display area, as illustrated in Fig. 8-8. Then press the down arrow until the highlight bar is located on

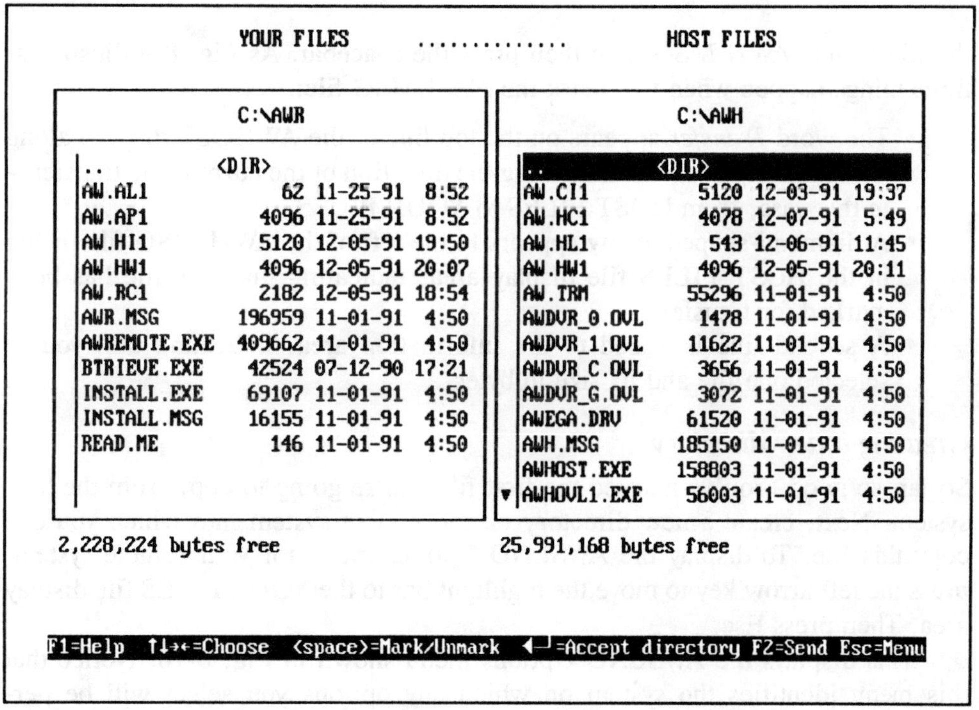

8-8 Use the arrow keys to pick the active file display.

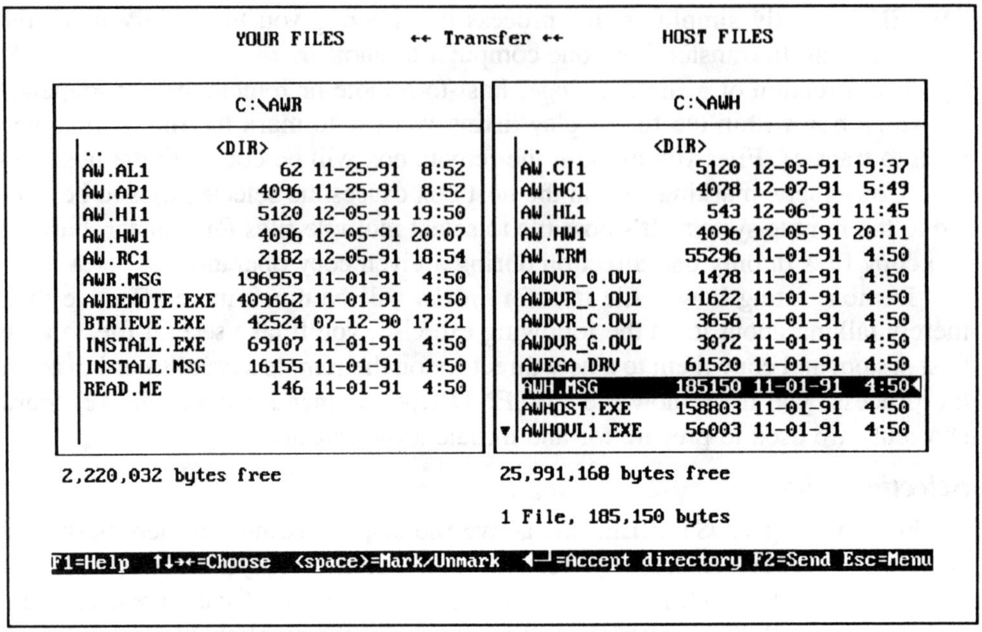

8-9 Use the spacebar to mark individual files.

the file named AWH.MSG, and then press the spacebar. As Fig. 8-9 illustrates, three things happen when you mark the AWH.MSG file:

- The word *Transfer* appears on the top line of the AWHOST display, along with a series of arrows indicating the direction of the current file transfer—in this case, from HOST FILES to YOUR FILES.
- A diamond-shaped arrow appears to the left of the AWH.MSG file listing in the HOST FILES file display area, indicating that this file has been marked for transfer.
- A second line is added to the information area, indicating that you've selected one file and its size in bytes.

Creating a new directory

So far, so good. You've marked the first file you're going to copy from the host system. Next, create a new directory on the remote system into which you can copy this file. To display the AWSEND Options menu for your remote system, press the left arrow key to move the highlight bar to the YOUR FILES file display area. Then press Esc.

This displays the AWSEND Options menu shown in Fig. 8-10. Notice that this menu identifies the system on which any options you select will be performed. In the case of Fig. 8-10, for example, the Options menu specifies YOUR FILES, or the remote computer, as the active disk drive. (Had the host system

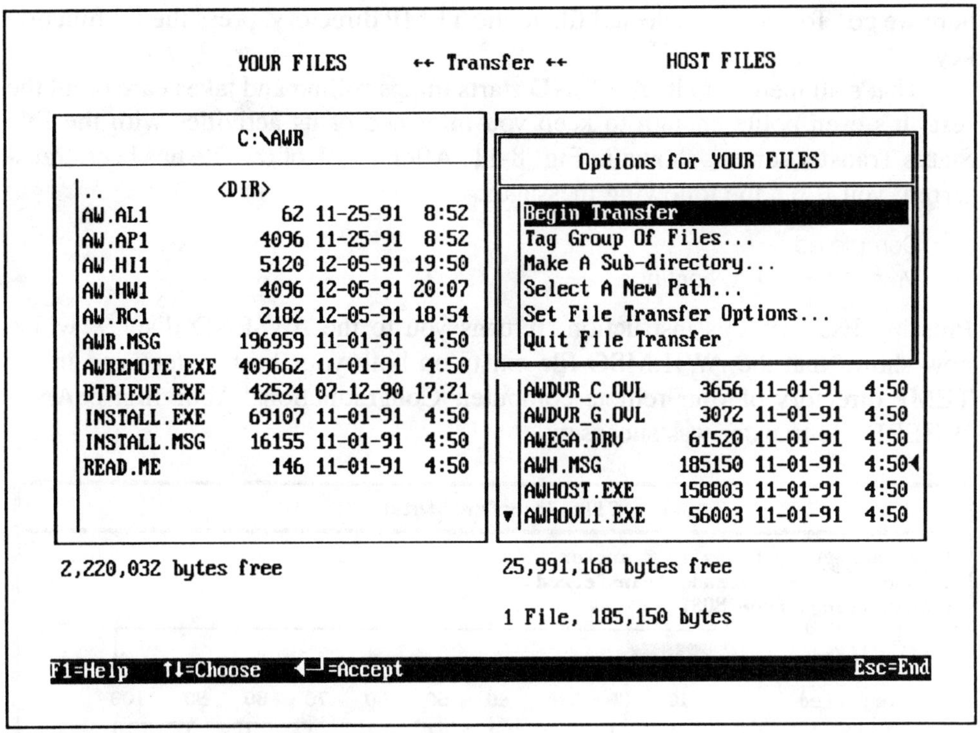

8-10 The AWSEND Options menu.

been active when you accessed the Options menu, its top line would have read *Options for HOST FILES*.) I'll describe all the AWSEND options in a few paragraphs. For now, let's go ahead and create the new directory. To create a new directory on the remote system:

1. Use the arrow keys to highlight Make a Sub-directory and press Enter.
2. AWSEND will display a prompt box, asking you to assign a name to the directory you want to create.
3. Type TEMP and press Enter.
4. Press Esc to remove the Options menu.

Changing active directories

The new directory now appears in the YOUR FILES file display area. To make this the active directory, press the down arrow to highlight the TEMP directory and press Enter. When you're returned to the AWSEND display, the new (and currently empty) TEMP will be the active directory in the display area.

Copying a file

Now that everything's set up, this is going to be so easy it hardly deserves a separate section. Still, you could make a valid argument for consistency in format. So here we go. To copy the selected file to the TEMP directory, press the F2 function key.

That's all there is to it. AWSEND starts things rolling and takes care of all the rest. It's even polite enough to keep you informed of its activities with the File Status Transfer screen, shown in Fig. 8-11. After 100% of the file has been transferred, you'll see the following message:

Completed
Press <Esc> to continue . . .

Pressing Esc, per this instruction, returns you to the AWSEND display, which now shows that the AWH.MSG file exists on both your host system and in the TEMP directory of your remote computer. Congratulations! Your first pcANYWHERE file transfer was successful.

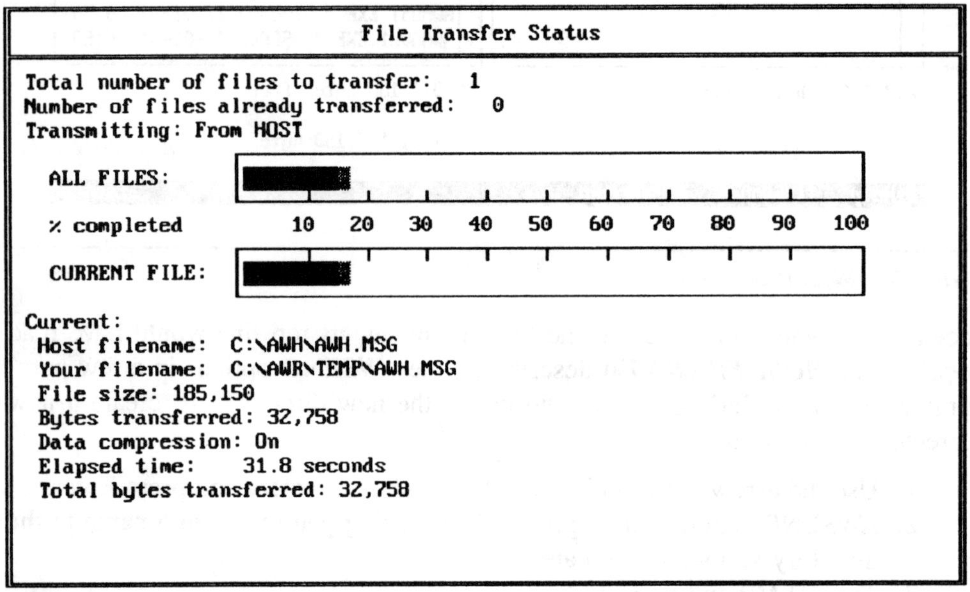

8-11 AWSEND's File Transfer Status display.

The AWSEND Options menu

Now that I've demonstrated AWSEND's simplest file-transfer method, I'm going to go back and fill in some gaps, specifically, some of the options available on the AWSEND Options menu. Before you begin, however, try the following:

1. If necessary, press the right arrow key to move the highlight bar to the HOST FILES file display area.
2. Press the down arrow key until the highlight bar is located once again on the AWH.MSG file.
3. Press the spacebar.

This once again marks the AWH.MSG file in your HOST FILES file display area. I had you do this because certain items are included on the AWSEND Options menu only when one or more files have been selected. Now when we display the Options menu, these items will also appear.

To display the AWSEND Options menu, press Esc. When the Options menu appears this time, as illustrated in Fig. 8-12, it will contain several new options in addition to those you saw previously (see Fig. 8-10). With all the options displayed, now seems as good a time as any to describe briefly what each one does.

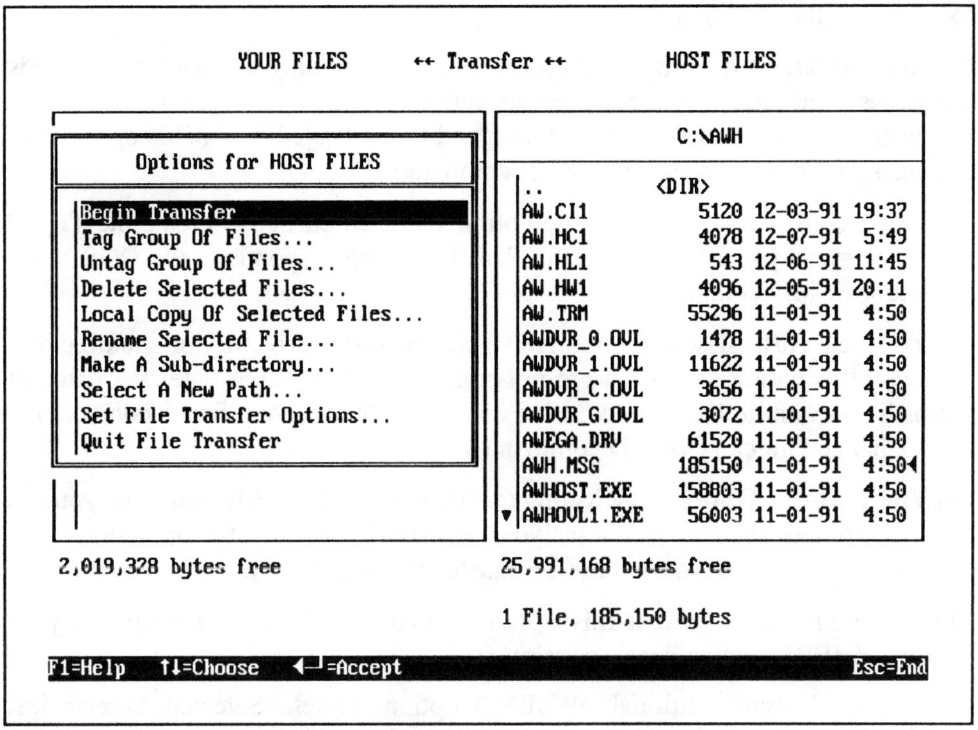

8-12 The Options menu displays additional items after a file is selected.

Begin Transfer

Selecting this option is the same as pressing the F2 function key. It begins transferring any marked files to the specified destination.

Tag Group of Files This option lets you quickly mark several files using the standard DOS wildcard characters (∗ and ?). More on the DOS wildcards in a few paragraphs.

Untag Group of Files This option also uses the DOS wildcards to identify multiple files, but removes the specified files from an already tagged group of files.

Delete Selected Files Selecting this option deletes any tagged files from the active directory. You're provided an opportunity to specify whether you want to verify each deletion before the file is actually erased. (This option corresponds to the DOS Delete or DEL command.)

Local Copy of Selected Files This option provides a method for copying a file to a different disk or directory on the same system, rather that to the other system in a remote connection. In other (and probably easier to understand) words, you would use this option to copy a file from one location on the host system to another disk or directory on the host system or, alternately, between two different locations on the remote system.

Rename Selected File AWSEND displays this option only when a single copy is tagged, as is the case now. Selecting this option displays a prompt box, requesting you to enter the new name you want assigned to the tagged file. (This option corresponds to the DOS Rename or REN command.)

Make a Sub-directory This is the option you used earlier to create the TEMP subdirectory on your remote system. (This option corresponds to the DOS Make Directory or MD command.)

Select a New Path Use this option to switch the active directory or subdirectory. Selecting this option displays a prompt box, requesting you to enter the name of the path you want to be the new active directory. (This option corresponds to the DOS Change Directory or CD command.)

Set File Transfer Options Selecting this option displays a submenu that you can then use to specify how you want your AWSEND file-transfer operations performed. I'll discuss the available file-transfer options shortly.

Quit File Transfer Select this option to exit AWSEND and return to your pcANYWHERE remote-access session.

Note: There's one additional AWSEND option, Delete Selected Directories, which appears on the Options menu only when you've tagged a directory, rather than a file or group of files, in the file-display area. If available, this option erases the tagged directory, all the files it contains, and any subdirectories running off of it with a single keystroke. I'll repeat here what the pcANYWHERE documentation says about this command:

USE THIS OPTION WITH EXTREME CAUTION!

That pretty well sums up my feelings. In other words, if you don't know exactly what you're doing, avoid the Delete Selected Directories option religiously—'nuff said? With your TEMP directory in place and a remote connection firmly established, let's try using some of the options to really put AWSEND through its paces.

Tagging multiple files

The key to having AWSEND tag several files at once is understanding how to use * and ?, the DOS wildcard characters. When included in a filename, an asterisk (*) indicates that any character can occupy that position or any of the remaining positions within a filename or its extension. When included in a filename, a question mark (?) indicates that any character can occupy that position only.

Basically, the asterisk replaces all characters either in a DOS filename up to the period separating the filename and extension, or in the filename's extension up to the end of the entire filename. The * wildcard tells DOS, in essence, to perform the current operation on any file(s) matching this partial description, regardless of the character(s) following the asterisk I have entered. A simple example of how you can use the asterisk with AWSEND's Tag Group of Files option will demonstrate how this works.

To tag all files ending with the extension DRV in your AWH directory, use the arrow keys to highlight Tag Group of Files and press Enter. AWSEND will display a prompt box, asking you to enter the pattern, including wildcard characters, that identifies the files you want to tag. Type *.DRV and press Enter.

When you return to the AWSEND display screen, you'll notice that, in addition to the AWH.MSG file, a second file, AWEGA.DRV, has been tagged. So you tagged one additional file. It certainly doesn't seem like you accomplished very much, does it? Before becoming too disappointed, however, try pressing PgDn twice. This last step demonstrates two important items:

- You can use the PgDn key to scroll through file listings too big to fit in a single file-display window.
- The DOS * wildcard character worked. Your screen should resemble Fig. 8-13, which clearly indicates that every file with a DRV extension has been tagged.

The ? wildcard differs from an asterisk in that it's used to replace a single character within either the filename or extension portion of a filename. To see how the ? wildcard works:

1. Press Esc to display the AWSEND Options menu.
2. Use the arrow keys to highlight Tag Group of Files and press Enter.
3. This time, when AWSEND displays its pattern prompt box, type AWVDD?A.* and press Enter.

This will tag two additional files within your HOST FILES file listing, AWVDDEGA.386 and AWVDDVGA.386. The only difference between the filenames is their sixth character, an E in the former and V in the latter. By substituting the DOS ? wildcard for this character, you've managed to tag both of these files at the same time.

Tip: As I rather slyly demonstrated in the previous exercise, it's perfectly

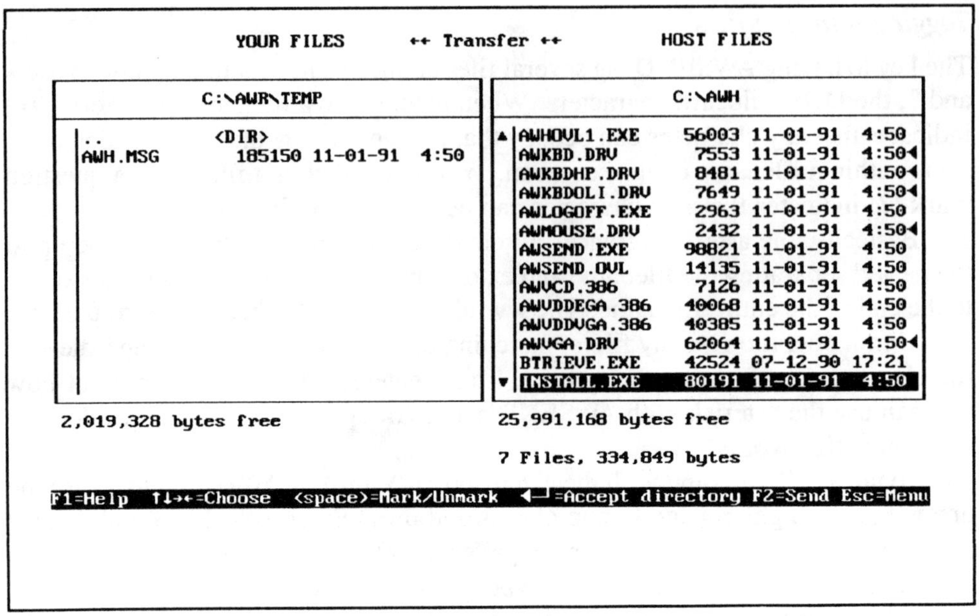

8-13 Pressing F3 copies the selected file to your remote computer.

legal to combine both DOS wildcards within a single AWSEND command. While this can get tricky, it does allow you to be quite explicit about the files you want to use with a given DOS operation.

The DOS wildcard characters also can be used with the Untag Group of Files option. The concept is precisely the same, only the results differ. Instead of adding files to your tagged list, the untag option removes any files matching the pattern you specify.

Setting transfer options

You have quite a few files marked in the AWH directory of the host system. Enough, in fact, that things could become confusing when the time comes to transfer them to the remote computer. Let's examine another AWSEND option, Set File Transfer Options, to see if pcANYWHERE offers any assistance to help you avoid some of this confusion.

To set transfer options, press Esc to display the AWSEND Options menu. Then use the arrow keys to highlight Set File Transfer Options and press Enter. This will display the AWSEND Preferences menu shown in Fig. 8-14, which lists the current settings that control how pcANYWHERE manages file transfers. Options contained on this list include:

Destination File Use this option to indicate what AWSEND should do when a file with the same name as one it's told to transfer already exists on the target system.

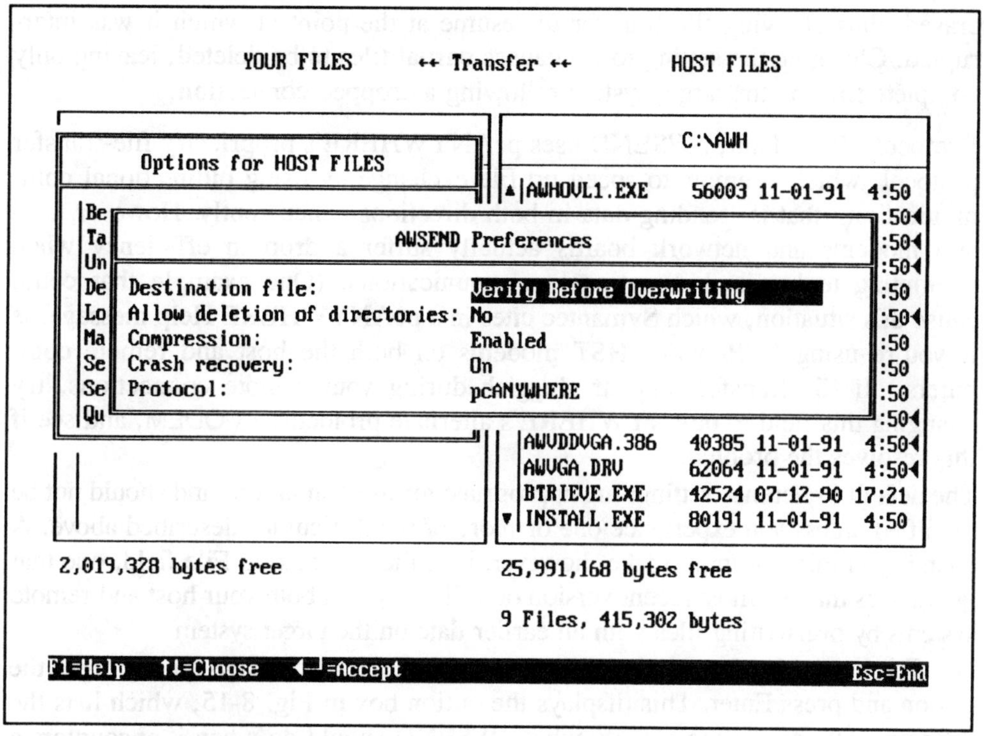

8-14 The AWSEND Preferences option box.

Allow Deletion of Directories This option relates to the caution I issued earlier. With this option set to No, my warning rings somewhat hollow, however, because AWSEND itself prevents you from performing wholesale directory deletions. Changing the default setting to Yes allows you to remove directories and subdirectories with the appropriate AWSEND option. (pcANYWHERE stores whatever setting you specify in your configuration file, so you won't need to modify it each time you run AWSEND.)

Compress If enabled, AWSEND compresses files to increase the speed of file exchanges. pcANYWHERE applies the highest level of compression it feels will work reliably, based on the type of connection established between the host and remote systems and the baud rate at which data is being transferred. If data is frequently garbled during file transfer, you might try changing this setting to Disabled. pcANYWHERE temporarily disables this feature when transferring files that have already been compressed using a file-compression program such as ARC and PKZIP.

Crash Recovery The setting in this field determines what AWSEND will do with files that aren't completely transferred when a pcANYWHERE connection fails during file exchange. If set to its default setting, On, partial files aren't erased, thus allowing file transfer to resume at the point at which it was interrupted. Changing the setting to Off causes partial files to be deleted, leaving only complete files on the target system following a dropped connection.

Protocol By default, AWSEND uses pcANYWHERE's proprietary file-transfer protocol, which attempts to speed up file exchange by using bidirectional communication—that is, sending data in both directions concurrently. However, certain modems and network boards actually suffer a drop in efficiency when attempting to handle bidirectional communications. (One example that could cause this situation, which Symantec cites in a pcANYWHERE Help message, is if you're using USRobotics HST modems on both the host and remote computers.) If file transfers appear sluggish during your remote connections, try changing this field to pcANYWHERE's alternate protocol, ZMODEM, and see if this resolves the problem.

The default Preference settings are appropriate for most situations and should not be modified unless you experience one or more of the difficulties described above. A change you might want to make, however, is in the Destination File field, one that guarantees that the most recent version of a file exists on both your host and remote systems by overwriting files with an earlier date on the target system.

To change the Destination File setting, use the arrow keys to highlight the option and press Enter. This displays the option box in Fig. 8-15, which lists the different settings used to specify what AWSEND should do when it encounters a preexisting copy of a file on the target system.

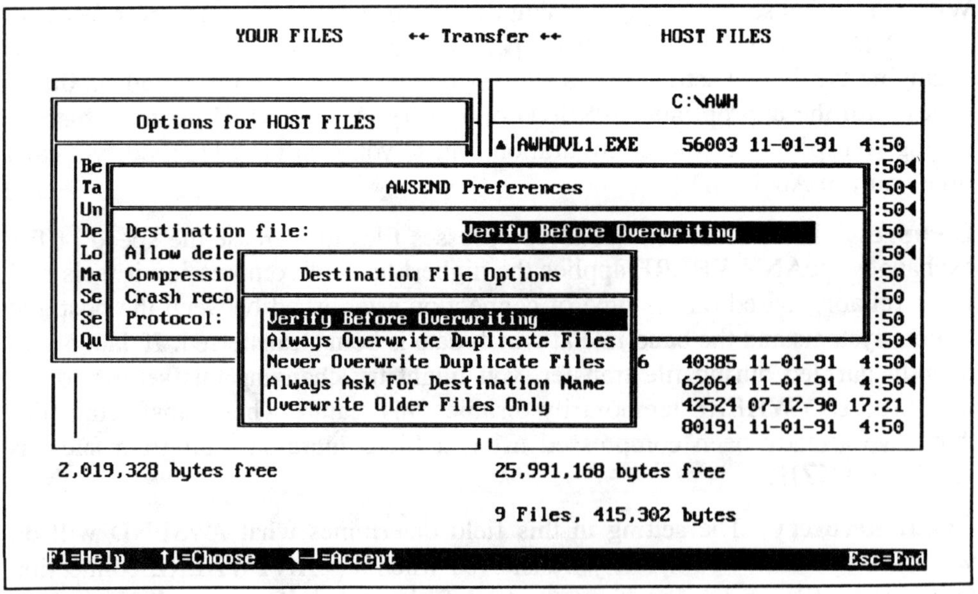

8-15 You can instruct AWSEND how to handle duplicate files it finds on the target disk.

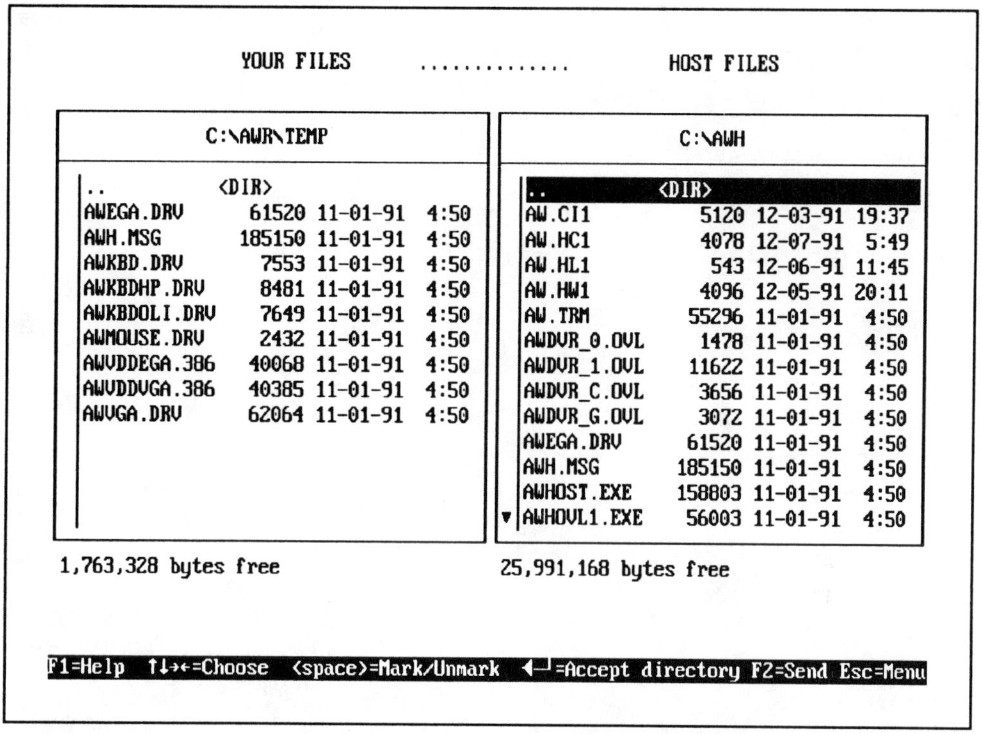

8-16 Selecting the Begin Transfer option copies all marked files to your remote computer.

To overwrite older files and return to the AWSEND Options menu, use the arrow keys to highlight Overwrite Older Files Only, press Enter, and then press Esc.

Now tell AWSEND to begin transferring your marked files using the current settings. To initiate the file transfer, use the arrow keys to highlight Begin Transfer and press Enter. Once again, AWSEND automates your file exchange. This time, however, because you're transferring multiple files, the percentages reflected in the ALL FILES and CURRENT FILE gauges on the File Transfer Status display (see Fig. 8-11) will differ from one another.

When the ALL FILES gauge reaches 100% and the Completed message appears, press Esc to return to the main AWSEND display. Your screen should now resemble Fig. 8-16, which shows that all the files tagged during the previous exercises have been successfully transferred to the TEMP directory created earlier. As these few examples illustrate, AWSEND provides a powerful tool for managing file transfers between your host and remote systems.

To exit AWSEND, press Esc to display the AWSEND Options menu. Use the arrow keys to highlight Quit File Transfer and press Enter. You're not quite finished with AWSEND. You'll encounter it again, briefly, in the next chapter as you explore individual options on the pcANYWHERE Sessions menu.

9
CHAPTER

Using the Sessions menu

Borrowing freely from an old baseball metaphor, you've just about covered all the critical bases when it comes to using pcANYWHERE for remote computing. This chapter, in fact, will be the last one in which you'll concentrate exclusively on that particular aspect of pcANYWHERE. In it, you'll examine how to use the various items on the Session Options menu, which were introduced and described briefly in the previous chapter.

As you might recall, I stated then that the Session menu provides you with a convenient method for utilizing the resources of your host and remote systems, placing the power of two computers literally at your fingertips during a remote-computing session. The results are almost gestalt in nature—creating a total PC environment that can, at times, exceed the sum of its individual parts.

The first step toward tapping this power is to access the Session Options menu. To do this, press Ctrl−Right Shift. Keep in mind that your host and remote systems must be involved in an active remote connection, with both machines running the appropriate pcANYWHERE module for you to access the Session Options menu, shown in Fig. 9-1. If you're using pcANYWHERE/Remote to contact a computer that isn't running pcANYWHERE/Host, you'll see a different menu containing what pcANYWHERE refers to as its *terminal options*. Because such connections fall under the heading of more traditional on-line activities, I'll defer any discussion of the terminal options menu until Part 3 of this book, *Beyond the basics*.

Now that the Session Options menu is available, start checking out what it allows you to accomplish. Tip: During installation, pcANYWHERE defines

9-1 The Session Options menu.

Ctrl–Right Shift as the default hot-key command used to display its Session Options menu. You can use the General option on pcANYWHERE/Remote's Preferences menu to select a different default hot-key combination, should this initial setting clash with specific programs you'll be running on the host system during a remote-computing session.

Communicating with a user at the host system

One useful feature of the Session menu is that it allows you to communicate directly with someone sitting at the host system over your pcANYWHERE connection. This ability can prove invaluable under certain circumstances.

Suppose, for instance, that you and a colleague are relying on pcANYWHERE to put together an important project. An example that immediately comes to mind would be a situation in which two people in two different locations must work with one another over a remote connection to refine a project budget using a popular spreadsheet program like Lotus 1-2-3.

Assumptions are a critical part of any budget projection. Consequently, you would want to know precisely what assumptions your coworker is applying as he begins plugging numbers into your shared spreadsheet. What simpler way to find out than by asking? You can, thanks to pcANYWHERE's Chat option.

Initiating an on-line chat

To initiate an on-line chat from the Session Options menu, use the arrow keys to highlight the Chat with Host User option and press Enter.

Figure 9-2 shows what you might see under such circumstances. The bottom portion of this screen contains an actual Lotus 1-2-3 spreadsheet that I'm running

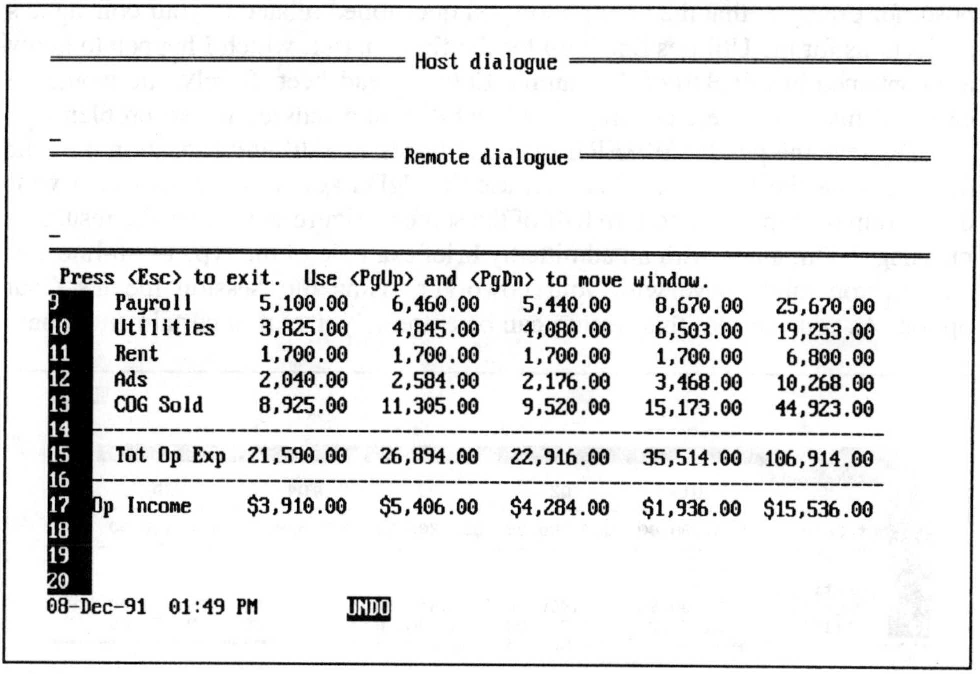

9-2 The pcANYWHERE Chat option lets you communicate directly with the host system.

on my host system. Notice, however, the two dialog boxes in the top half of this figure. These boxes, which appear whenever you activate the pcANYWHERE Chat feature, provide a way for you and the host user to exchange comments, questions, suggestions, and so forth with one another during an on-line session. Anything you type at your keyboard appears in the lower box, the one marked Remote Dialogue. Comments typed in at the host system appear in the upper portion of this display, marked Host Dialogue.

Note: As I pointed out in the previous chapter, I have no way of knowing the precise software you have installed on your host system. Although I'll be using Lotus 1-2-3 in the exercises that follow, there's no guarantee the you have access to this popular PC program. That's okay. The concepts I'll be discussing can be applied to almost any pcANYWHERE remote-computing session. So why not just go ahead and load one of your favorite applications on your host system, then follow along as best you can? Doing so will allow you to experiment with the various Session options. Just keep in mind that what you see on your display will differ from the figures here.

Moving the dialogue boxes

pcANYWHERE even takes precautions to ensure that it doesn't obscure information displayed on-screen that you and the host user might need to discuss. Suppose, for example, that the assumption you questioned related to your colleague's projections for the Utilities line item for the first quarter, which I happen to know are contained in cell B10 of the sample Lotus spreadsheet. Surely, life would be easier if this figure were actually visible while you discussed it. No problem.

To move the pcANYWHERE dialogue boxes to a different location on your display, press the PgDn key. You can use the PgDn key to move your chat windows from the top to the bottom half of the screen. Figure 9-3 shows the results of pressing PgDn, along with an admittedly brief example of the type of on-line discussion you might have with your coworker, using the Session menu's Chat option. As you can see, this feature can be extremely useful in certain situations.

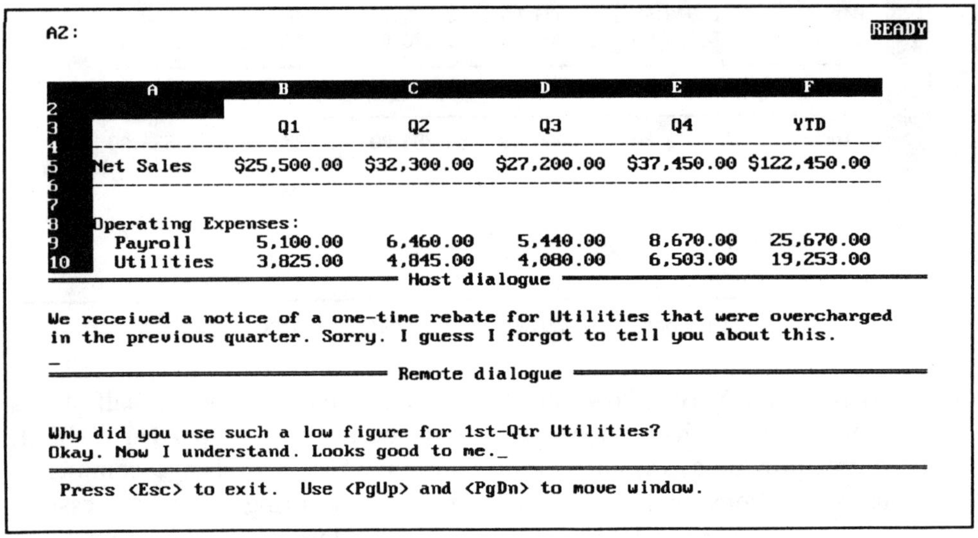

9-3 You can use the PgDn key to move the Chat window.

Ending an on-line chat

Of course, like all good conversations, your on-line discussion must end sometime. Let's terminate this one now by closing down the Chat session. To exit, press Esc.

This ends the on-line Chat session and returns you to the application program that was running previously on the host system. While I have it all set up, I might as well continue using the sample Lotus 1-2-3 spreadsheet to demonstrate some additional options on the pcANYWHERE Session menu.

Capturing screens from the host system

Let's suppose that, after resolving your question regarding that Utilities entry, you decide you'd like to have a copy of this portion of the projected budget to study later. You could print one out using pcANYWHERE's Print Options. An easier and faster approach, however, might be to take a "snapshot" of the host display and store it in a disk file on your remote system. This, too, is easy to do, thanks to another Session option, pcANYWHERE's Save Screen feature.

To save the current screen, press Ctrl–Right Shift to redisplay the Session menu, use the arrow keys to highlight Save Screen, and press Enter. This will display the prompt box shown in Fig. 9-4, asking you to specify the path (that is, disk drive and directory name) in which you want your screen snapshot stored.

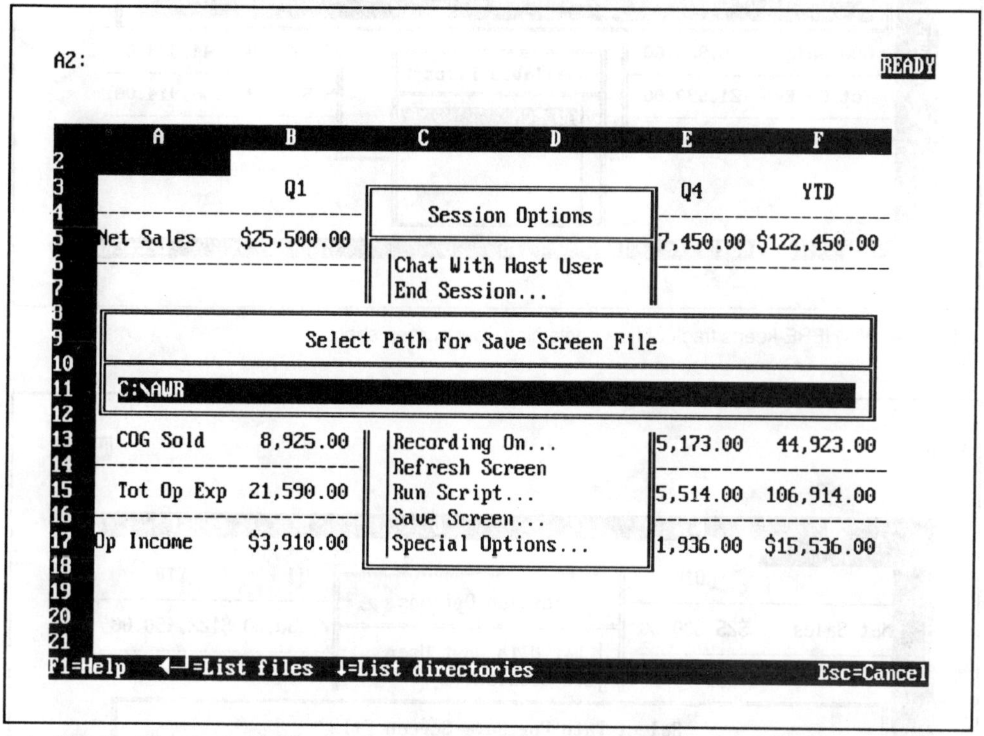

9-4 Use the Save Screen option to take a snapshot of the host display.

To accept the default directory, AWR, on drive C of your remote system, press Enter. You'll then be asked to specify a name for the file in which you want to store this screen. As Fig. 9-5 illustrated, pcANYWHERE creates a generic screen file on your remote system during installation and names it AW.SCN. Simply pressing Enter at this point would append the current screen to—that is, add it to the end of—the AW.SCN file. Instead, however, create a new file for your budget figures.

To create a new screen file, Press the Ins key to select New File. At the subsequent prompt, shown in Fig. 9-6, use the backspace key to erase AW, then type BUDGET and press Enter. pcANYWHERE will save the current contents of the

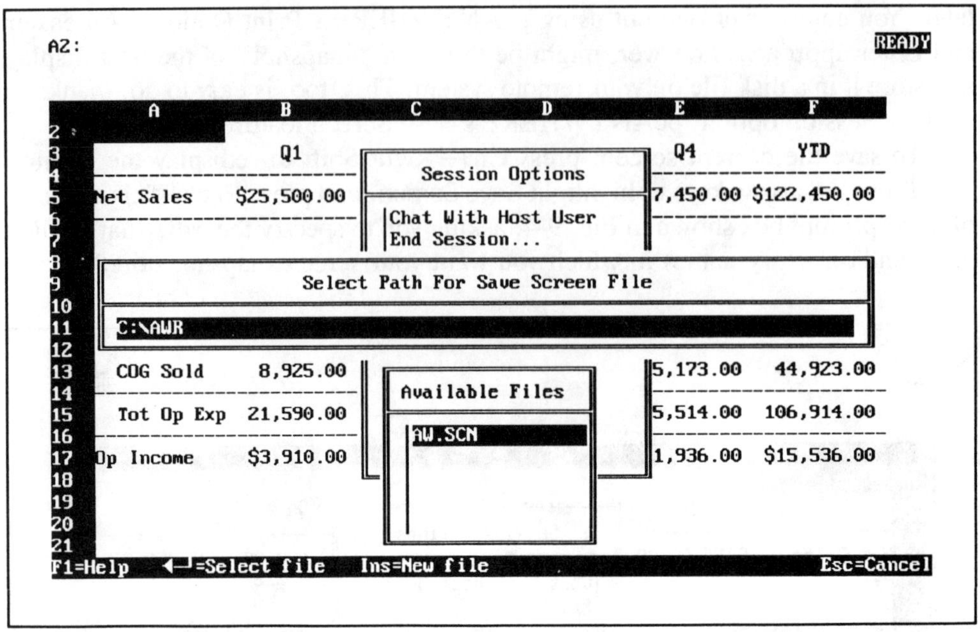

9-5 pcANYWHERE keeps track of files containing your screen shots.

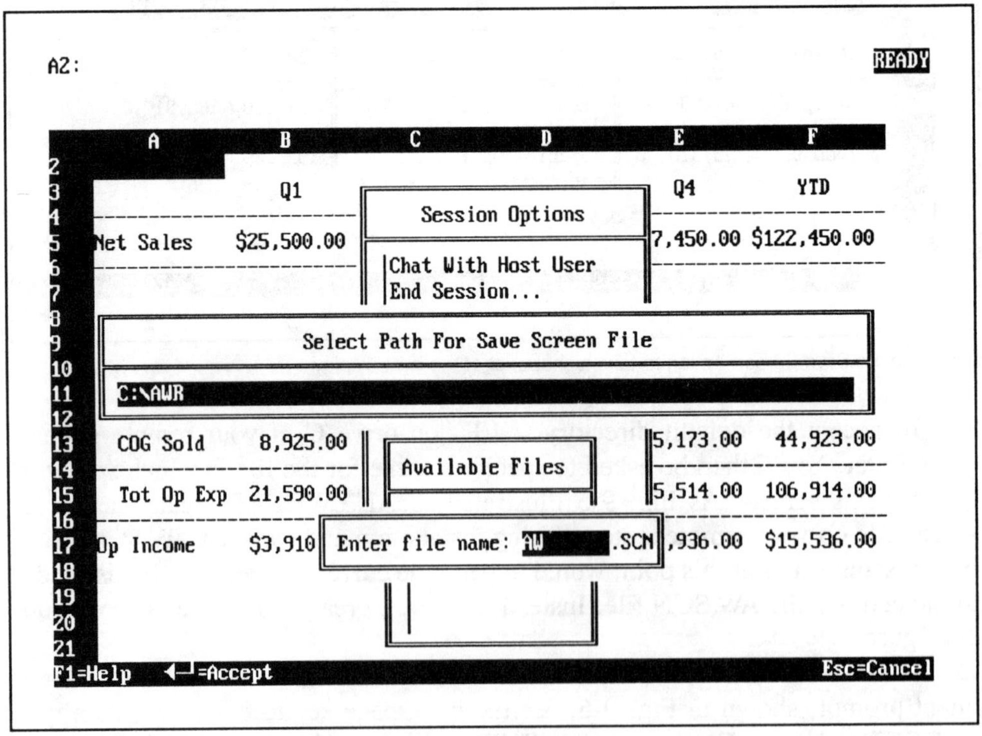

9-6 pcANYWHERE lets you specify where you want to store the screen shot.

host screen to the specified file and display the following message:

> The screen has been saved.
> Press <Esc> to continue . . .

Sounds like good advice to me. So press Esc. This will return you, once again, to the sample Lotus spreadsheet, which is still running on my host system.

Would you like to verify that pcANYWHERE indeed captured this screen? Given that your remote computer is dedicated to controlling the host system, you might think that this would be impossible until you exit pcANYWHERE/Remote. It isn't. Remember how I told you earlier that you actually have access to the resources of both your host and remote computers during a pcANYWHERE session? It's finally time to demonstrate what I mean by this.

Doing DOS with pcANYWHERE/Remote

One item on the Session Options menu actually lets you directly control the remote computer while still maintaining your on-line connection to the host system. Let's see how this works. To take direct control of your remote computer, press Ctrl – Right Shift to redisplay the Session menu. Use the arrow keys to highlight the Go to DOS option and press Enter.

Your display should now resemble Fig. 9-7. As this screen illustrates, the Go to DOS option temporarily suspends a remote session and opens a shell to the DOS operating system, from which you can execute DOS commands on your remote computer. (You can tell I'm accessing DOS on the remote computer because the system prompt in Fig. 9-7 lists a directory called AWR. You might recall that this is where I installed my copy of pcANYWHERE/Remote, way back in chapter 4.) While you're here, why not verify that pcANYWHERE did indeed capture your host display earlier?

To see a listing of screen-shot files in your AWR directory, type DIR *.SCN and press Enter. If all went as planned, you should see two filenames in the result-

```
Enter EXIT to return to pcANYWHERE


Microsoft(R) MS-DOS(R) Version 5.00
             (C)Copyright Microsoft Corp 1981-1991.

(C:\AWR)
```

9-7 pcANYWHERE includes a DOS shell that lets you access the resource on your remote computer during a remote session.

ing directory listing, AW.SCN and BUDGET.SCN. The latter is, of course, the file you created in the previous exercise.

Tip: pcANYWHERE/Remote even provides a special utility that allows you to view any screens you capture during a remote session. I can't cover everything at once, unfortunately, so, with your indulgence, I'll delay the discussion of this particular feature until the appropriate time—specifically, chapter 13, *The pcANYWHERE utilities*.

As this relatively simple demonstration illustrates, the Go to DOS option gives you quick access to all of the resources on your remote computer, even if you're already involved in a remote-access session. Issuing the appropriate DOS command to run an application program installed on the remote system is no more difficult than entering the DIR command was, but remember one important thing—to maintain your remote connection, pcANYWHERE must leave a portion of itself resident within memory, thus forcing you to sacrifice valuable RAM. Consequently, you might discover that certain memory-hungry programs that work just fine on your remote computer when you don't have pcANYWHERE/Remote running will be too large to execute from within the pcANYWHERE DOS shell.

Note: Depending on what type of computer your remote system is, you might be able to circumvent this problem by loading the pcANYWHERE/Remote software in a special section of memory called high RAM. I'll explain how you accomplish this, along with providing potentially useful suggestions on several other ways you can "tweak" your pcANYWHERE environment, in chapter 14, *Tips, techniques, and troubleshooting*. (Don't you just love the way I keep teasing you with future discussions? It's kind of like the "Previews of Coming Attractions" in movie theaters, isn't it?) Keep in mind that there are just too many facets to pcANYWHERE to cover at one time. I promise we'll get to everything.

As I've done before, I'll recommend that you experiment with the pcANYWHERE DOS shell. Try loading a few of your favorite programs and see if they run properly. Unless you experience the problem relating to the memory shortage I mentioned earlier, they should provide you with a concrete example of what I mean when I say that the DOS shell gives you full access to the resources of both your host and remote systems during a pcANYWHERE session. When you're ready to get back to remote computing, type EXIT and press Enter to leave the DOS shell and continue remote computing.

Setting the Special Options

This time, rather than taking you all the way back to the application program running on your host system, pcANYWHERE returns you to the Session Options menu. That's fine, because you still have one very special option to explore—called, in fact, the Special Options option—before returning to the host system. To

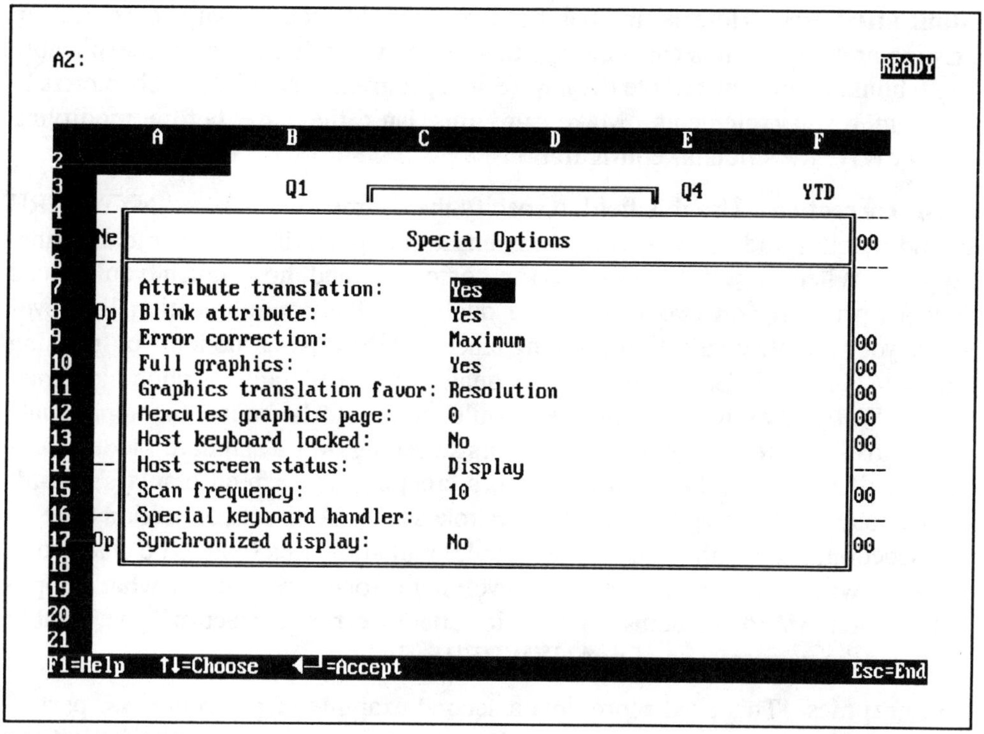

9-8 The Special Options dialog box.

select Special Options, use the arrow keys to highlight Special Options and press Enter.

As mentioned in the previous chapter, selecting this option will display a dialog box, shown in Fig. 9-8, which contains several settings that influence how the host and remote computers interact with one another. As I pointed out then, for example, it might be necessary to modify certain screen attributes when a remote computer with a monochrome display is used in conjunction with application programs set up to run in color on the host. If you've experienced no difficulties up until now, you might never need to modify the default settings pcANYWHERE assigns to these fields during installation. Still, understanding how each of the Special Options are used could prove essential should you encounter a problem during some future session. Program settings available in the Special Options dialog box include:

Attribute translation Changing this setting to No can reconcile potential compatibility problems between a color host and a remote system that uses a monochrome, or so-called Hercules-compatible, video adapter. Not all monochrome systems require modifying this setting. It should be changed only if you experience display problems during a pcANYWHERE session.

Blink attributes Here is another field you should modify only to correct an existing problem. In this case, change this setting to No if characters inexplicably begin blinking on your remote display. (Some programs use blinking characters to highlight display elements. Make sure this isn't the case before modifying pcANYWHERE's default configuration.)

Error correction Use this field to specify the degree to which pcANYWHERE should monitor and, if necessary, compensate for potential errors in data transmission. Whether you'll require error correction and how much you'll need depends primarily on two factors, the quality of the communication link over which your remote connection is taking place and the type of modem you're using with pcANYWHERE. Given the poor quality of today's analog phone systems, pcANYWHERE assigns this field a default setting of Maximum, the appropriate setting to eliminate the possibility of errors surfacing over a standard modem connection. If the possibility of minor errors creeping into screen displays doesn't bother you, try changing this field to Keystrokes Only. A direct connection (that is, a connection made with a null-modem cable) will allow this field to be changed to None, as will phone-line connections involving error-correcting (or what's popularly called *MNP*) modems. When in effect, error correction unavoidably degrades the performance of pcANYWHERE slightly.

Full graphics This field represents a second example of perfection vs. performance. When Full graphics is set to Yes, its default setting, pcANYWHERE retransmits a complete image to the remote system whenever the host display changes. This ensures an exact reproduction of host screens on the remote system. However, it also tends to slow down pcANYWHERE due to the time required to pass such comprehensive visual data back and forth over a remote connection. As was the case with error correction, if minor errors in screen displays won't bother you, setting this field to No will improve pcANYWHERE's performance.

Graphics translation favor The options for this field are Color and Resolution. The default setting, Resolution, matches the quality of the host and remote displays to one another as closely as possible while still reproducing colors accurately. It's the better setting when an EGA or higher resolution monitor is being used on both the host and remote systems. If you have a CGA or Hercules-compatible display on your remote computer, changing this setting to Color might resolve potential problems relating to color reproduction, but at the expense of display clarity.

Hercules graphics page You need to be concerned about this field only if your remote system uses a Hercules-compatible monochrome display. (In fact, pcANYWHERE makes this field "off limits" to you if your default configuration specifies anything other than a monochrome display.) Because the Hercules standard uses multiple display pages, it might be necessary to modify the setting in this field to ensure that host screens are being transmitted properly to a monochrome remote system.

Host keyboard locked Changing this field to Yes disables the keyboard on the host system. Doing so can be useful when you want to keep anyone with access to the host system from interfering with your remote activities. This might be the case, for example, if you're performing system diagnostics remotely to find and correct problems on the host system. People who use pcANYWHERE to provide remote training might also want to take advantage of this feature because it allows them to determine if and when the persons they're instructing can interact with the host system.

Host screen status Use this field to blank out the display on the host system, a useful security feature if you're working with sensitive material over a remote connection.

Scan frequency This field specifies how often pcANYWHERE checks the host PC screen to see if display changes need to be transmitted to the remote computer. Because the value in this field is specified in increments of 55 milliseconds, the default setting (10) indicates 550 milliseconds. If your remote display jiggles or jumps, try changing this setting to a lower value. Doing so might correct the problem, but will cause the host PC to run more slowly.

Special keyboard handler This feature, which is described in chapter 5, can be enabled or disabled from the host system only. If special keyboard handling is enabled at the host, this field will show its current setting.

Synchronized display When set to Yes, the default setting, this field causes the displays on the host and remote system to always match one another. In other words, each change in the host display is reflected on the remote computer. While this normally presents no problem, it can slow down a remote session if you're doing something that causes the display on the host system to be updated continuously, as would be the case, for example, when scrolling through a long document while running a word processor. Changing this setting to No causes the screen on the remote computer to be refreshed only after activities on the host system have stopped and its display has stabilized, thus increasing the speed of a remote session when you're using certain applications.

As I stated at the beginning of this section, chances are you'll never need to modify any of the special settings unless the default values cause problems during a remote session or for security purposes, as outlined above. Of course, you might want to "tweak" your environment slightly to improve the overall performance of pcANYWHERE, in which case the above explanations will help you know which Special Options to change and how to adjust them. To exit the Special Options dialog box, press Esc.

Well, I guess it's time to follow through on a promise I made at the end of the previous chapter, when I told you you'd be returning shortly (and briefly) to the AWSEND file-transfer program. So, let's do it, shall we? But first, you'll need to be able to use the remote computer to access DOS on the host system. To access DOS on the host system:

1. Press Esc to leave the Session Options menu.
2. Close down any application program running on your host computer and return to the DOS system prompt.
3. Type C: and press Enter to make drive C your active disk drive.
4. Type CD \AWH and press Enter to make AWH the active directory.
5. Type CLS and press Enter to clear the screen.

Note: Blame this final step on my predilection for (some might call it an obsession with) neatness. I generally prefer to start off new exercises with a clean screen.

Running AWSEND from the DOS command line

AWSEND's menu-driven structure is extremely convenient for neophyte PC users. In this respect, it's not that different from Windows graphics-based environment or the DOS Shell included with newer releases of MS-DOS. Some people, however, prefer to "work without a net," so to speak, and enter commands directly at the DOS system prompt. Doing so is generally faster than dealing with an interactive environment, especially for "quick-and-dirty" operations.

Such folks will appreciate the fact that AWSEND allows you to initiate file transfers from the DOS system prompt. When run in this manner, AWSEND lets you transfer files between a host and remote system using a simplified format based on the same command syntax as the standard DOS COPY command. The proper syntax for an AWSEND command is:

AWSEND [*filename*] to|from host|remote [*pathname*]

Hmmmm. This doesn't look like a very "simplified format," does it? I'll tell you what. Let me instead provide a specific example that should clarify matters considerably and, in the process, convince you that the AWSEND command really is easy to use. Try typing the following:

AWSEND READ.ME TO REMOTE C:\AWR\TEMP

and pressing Enter. This command tells AWSEND to transfer a file called READ.ME to the remote system and place it in a directory called \AWR\TEMP, the work directory you created in the previous chapter. If you read that previous sentence carefully, and then look back at the above AWSEND command syntax, you should have no trouble figuring out how it all fits together.

As it did during your previous file transfers, AWSEND displays its File Transfer Status screen while the specified operation takes place. This time, however, you won't need to press Esc to remove that display after the file is transferred. Instead, AWSEND returns you automatically to the DOS system prompt. It's all quite fast and requires little effort. And when you're finished, what happened? To determine this:

1. Press Ctrl – Right Shift to redisplay the Session menu.
2. Use the arrow keys to highlight the Go to DOS option and press Enter.
3. Type CD \AWR\TEMP and press Enter to make TEMP the active directory.
4. Type DIR READ.ME and press Enter.

Notice that a copy of the READ.ME file now exists in the TEMP directory of your remote computer, as illustrated in Fig. 9-9. Clearly, the AWSEND command worked exactly as planned.

```
Enter EXIT to return to pcANYWHERE


Microsoft(R) MS-DOS(R) Version 5.00
             (C)Copyright Microsoft Corp 1981-1991.

(C:\AWR) CD \AWR\TEMP

(C:\AWR\TEMP) DIR READ.ME

 Volume in drive C is MS-DOS_5
 Volume Serial Number is 1719-6011
 Directory of C:\AWR\TEMP

READ     ME         4792 11-01-91   4:50a
        1 file(s)       4792 bytes
                     1775616 bytes free

(C:\AWR\TEMP)
```

9-9 AWSEND automatically copied the specified file to your TEMP directory.

Note: As it did when you were using the menu-driven display, AWSEND recognizes the DOS wildcards when it's run directly from the system prompt. For example, the above file could have been copied with the following command:

AWSEND READ.* TO REMOTE C:\AWR\TEMP

(Of course, this command would also have copied all other files that started with READ, regardless of their extensions, had any existed in the AWH directory of the host system.)

As I stated above, running AWSEND from the DOS command line provides a very efficient method for exchanging files between your host and remote systems. If you're new to the wonderful world of DOS, however, it might be wise to stick with AWSEND's interactive display until you become more familiar with how DOS commands operate.

Ending a remote session

The only thing left to do is to terminate the current remote-computing session and return to the pcANYWHERE/Remote Main Menu. So let's do it. To end the current remote session, type EXIT and press Enter to exit pcANYWHERE's DOS shell. Then use the arrow keys to highlight the End Session option and press Enter.

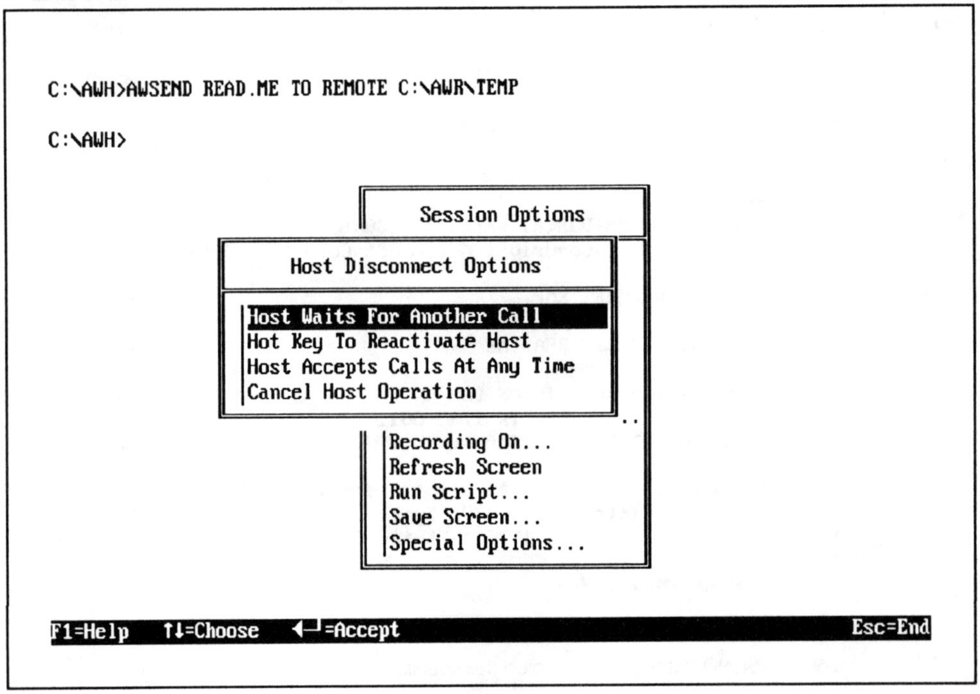

9-10 You reset the host system each time you end a remote session.

pcANYWHERE will display the option box shown in Fig. 9-10, asking how you want the host system configured after it disconnects the current call. The available settings, which correspond to similar items in the Begin Host Operations option box described in chapter 6, include the following:

Host Waits for Another Call Choosing this option keeps pcANYWHERE/Host active and waiting for the next call, following termination of the current session.

Hot Key to Reactivate Host Choosing this option returns your host system to the DOS system prompt, but leaves pcANYWHERE/Host running as a terminate-and-stay-resident (TSR) program. Someone at the host system will need to enter the appropriate hot-key command before pcANYWHERE/Host will accept an incoming call.

Host Accepts Calls at Any Time Choosing this option also sets up pcANYWHERE/Host as a TSR and returns the host system to the DOS system prompt. In this case, however, pcANYWHERE/Host remains active in the background, ready to automatically answer incoming calls from a remote system.

Cancel Host Operation This option causes pcANYWHERE/Host to attempt to remove itself from memory and return your host system to the DOS system prompt. After selecting this option, the only way to initiate a remote session is to have someone at the host system reload pcANYWHERE/Host into memory.

To terminate this sample remote session, press Enter to specify Host Waits for Another Call.

Saving or discarding a session log

The next screen you'll see on your remote computer should be the dialog box shown in Fig. 9-11. You might recall that you configured the Caller Information file for this remote session to automatically include session logging. Consequently, pcANYWHERE/Remote asks for a filename under which it should save the current log prior to returning you to the Main Menu. If you really want to save this series of exercises, feel free to do so by entering a filename of your choice at

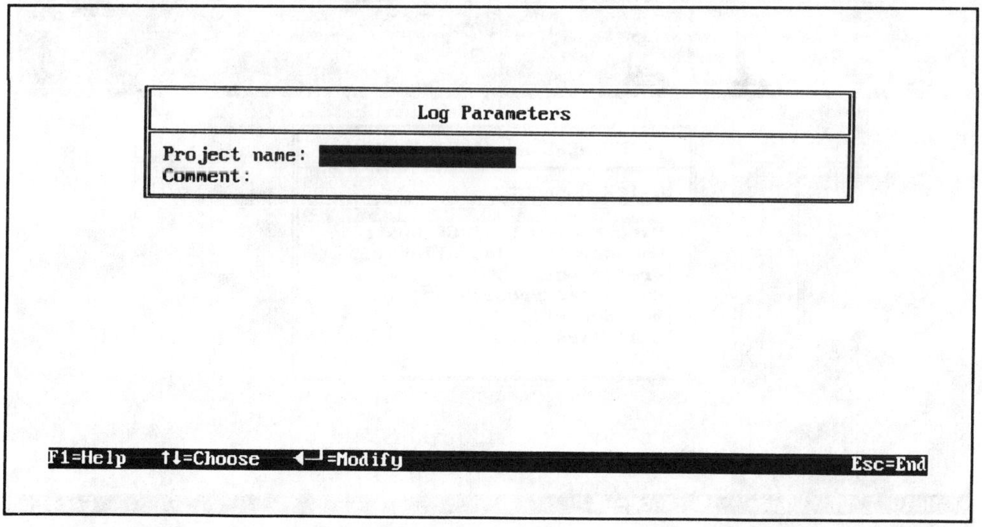

9-11 You can name and describe your log file.

the prompt screen. You can supply a more detailed description of your log file at the optional Comment field.

After entering this information, if desired, press Esc. This will display the Log Entry Options menu shown in Fig. 9-12. Again, whether you save or discard the current session is entirely your choice. Realize, however, that log records consume valuable disk space, so I'd recommend that you save them only when there's a logical reason for doing so. In the case of these initial exercises, I'd have to say there isn't.

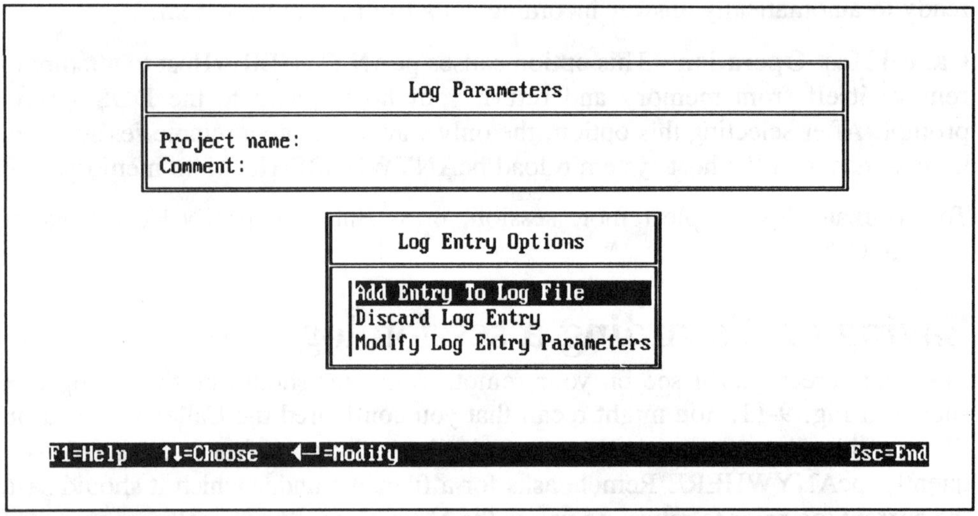

9-12 pcANYWHERE lets you decide whether to save or discard individual log entries.

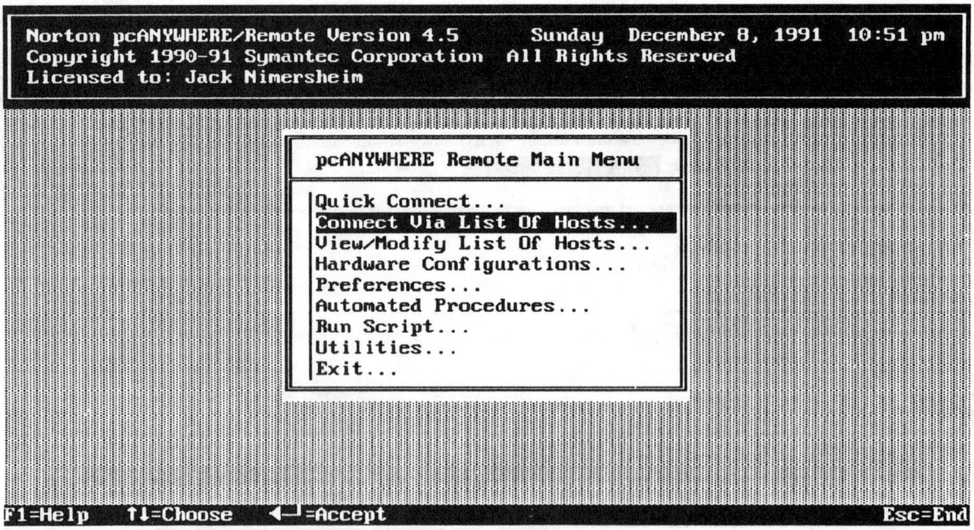

9-13 The pcANYWHERE/Remote Main Menu.

To discard this log entry, use the arrow keys to highlight the Discard Log Entry option and press Enter. Viola! This returns you to the pcANYWHERE/Remote Main Menu, shown in Fig. 9-13. Congratulations! You've just completed your first successful connection to a host system!

Summary

As I've stated several times throughout the previous eight chapters, remote computing is pcANYWHERE's primary *raison d'etre*. It's also a topic covered quite extensively, and with remarkable clarity, in the pcANYWHERE User's Guide. Consequently, I'm going to largely ignore this basic aspect of pcANYWHERE throughout the remaining chapters in this book. Rather, I'm going to move (as the title of Part 3 implies) *beyond the basics*, as we begin exploring the myriad of ways in which you can use pcANYWHERE for more traditional on-line activities. So take a deep breath. It's time to spread your electronic wings and really take flight!

10
CHAPTER

Using pcANYWHERE/ Remote to contact other on-line services

I have a confession to make: I love telecommunications. Given the opportunity, I could ramble on for pages about the enjoyment inherent in being able to use my PC, a modem, and a telecommunications program to access another system in another part of the world at virtually any time of the day or night. In fact, I write two monthly columns on telecommunications in which I do precisely that. Of all the PC applications I work with (including word processing, which one would think I'd be extremely partial to) telecommunications is my hands-down favorite. And if facts and figures count for anything, I'm not alone in my excitement.

How popular is telecommunications? Popular enough that CompuServe, only one of a dozen or so commercial information services currently doing business in this country, claims close to 1 million subscribers. During any given day, one recent survey estimated, anywhere from eight to fifteen million people use their computer for some kind of telecommunications-related task—everything from exchanging electronic correspondence, or e-mail, to ordering airline tickets and making hotel reservations for their latest business trip. By way of comparison, consider the following fact: Today, more than a decade after the first technical shots were fired in the so-called PC revolution, the most popular computer magazine on the newsstands boasts a total circulation figure that falls considerably short of the 750,000 mark.

Clearly, the siren song of telecommunications is heard far beyond the borders of the core community of PC users. And well it should be! As I pointed out all the way back in chapter 1, telecommunications—and, by extension, pcANYWHERE—places the entire world quite literally at your fingertips.

In this chapter you'll cast off our digital moorings and set about exploring the electronic nooks and crannies of the world. Prepare yourself for a wondrous journey, as you begin examining how, in addition to its remote-computing capabilities, pcANYWHERE/Remote can be used to perform a wide range of more traditional on-line activities. (This last observation assumes, of course, that the word *traditional* can be legitimately applied to an industry that's barely ten years old.)

pcANYWHERE's Quick Connect option

The secret to taking full advantage of pcANYWHERE/Remote's most impressive capability is, ironically, one of the program's least assuming features: the Quick Connect option. As its name implies, Quick Connect lets you set up your computer quickly for remote operations. But Quick Connect's usefulness goes far beyond being a "quick and dirty" method for getting on-line.

Unlike pcANYWHERE/Remote's host list, which defines specific communication parameters and even dials the phone for you, Quick Connect includes a Terminal option that, as you'll see later in this chapter, puts you in direct communication with your modem. In short, Quick Connect lets you begin an on-line session with a clean slate, so to speak, and then grants you near-total control over your telecommunications environment.

Selecting Quick Connect

Quick Connect is one of the options on pcANYWHERE/Remote's Main Menu. To select Quick Connect, use the arrow keys to highlight Quick Connect and press Enter. pcANYWHERE/Remote will display its Quick Connect screen, shown in Fig. 10-1. This screen consists of two elements: the Quick Connect Parameters dialog box and the Quick Connect Options menu. Let's see how each of these items is used, during a Quick Connect session.

Setting Quick Connect parameters

Figure 10-1 shows the Quick Connect screen as it appears immediately following installation, before any of its parameters have been modified to match your PC

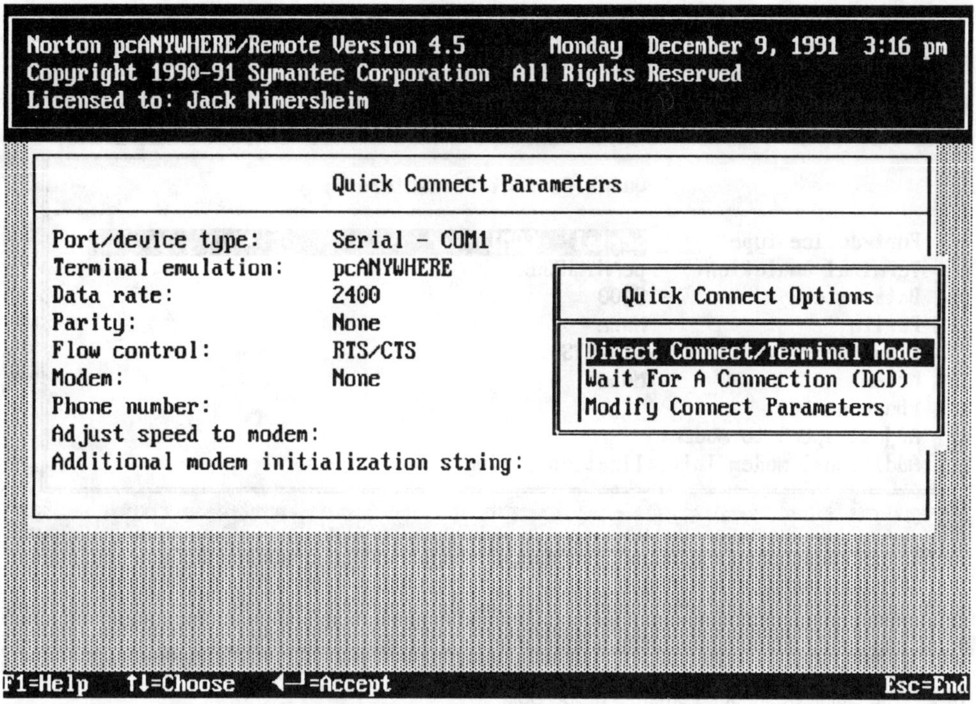

10-1 The Quick Connect Options menu.

hardware. Let's change these parameters first and see what influence it has on Quick Connect. To configure your Quick Connect environment, use the arrow keys to highlight Modify Connect Parameters and press Enter.

The Options menu disappears and Quick Connect advances you to its Parameters dialog box, as illustrated in Fig. 10-2. Modifying the Quick Connect parameters involves the same procedures that you used to configure your Caller Information files and Host List in earlier chapters—that is, you highlight the field you want to change and then select the appropriate setting for your hardware or, alternately, another computer you plan on calling with Quick Connect.

One way to use Quick Connect is to set it up to let you quickly contact the host system you anticipate you'll be contacting most frequently. Creating the appropriate entry in your pcANYWHERE/Remote Host List, however, will let you accomplish this just as easily—and just as quickly, despite the implied connotations of Quick Connect's official sobriquet. My suggestion, therefore, is that you configure Quick Connect to function as a generic on-line environment, a menu option you can choose in order to contact any one of several remote systems, depending upon which phone number you specify after selecting Quick Connect.

Note: I can see we're going to start tripping over terminology soon, not an uncommon occurrence in matters pertaining to computers, personal or otherwise. In the pcANYWHERE vernacular, *host* and *remote* are fairly straightforward

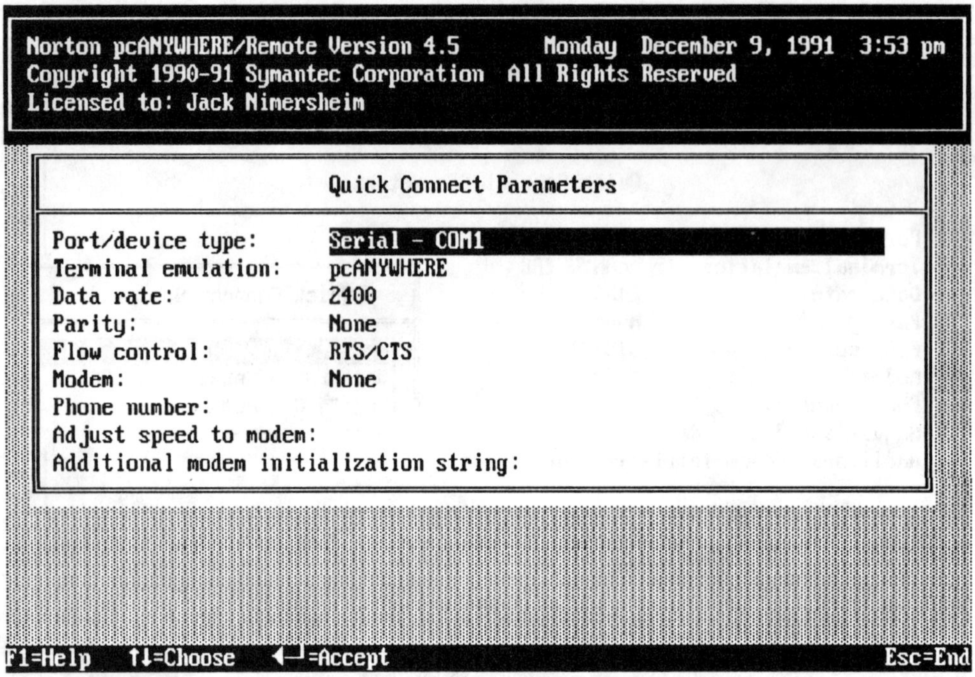

10-2 The Quick Connect Parameters dialog box.

terms. The host system is the computer on which you install and run pcANYWHERE/Host, while the remote computer is dedicated to pcANYWHERE/Remote. The latter system's sole purpose in a pcANYWHERE/Host-to-pcANYWHERE/Remote connection is to remotely take over and control the former, so the potential for confusion remains relatively minor. Move beyond using pcANYWHERE exclusively for remote computing, however, and the waters turn a little more murky, in that *remote* becomes a relative term. Wouldn't it be just as logical to refer to the computer you're calling (i.e., a computer in a remote location) as the *remote* system? It would, and many people do (including me) on many occasions. Consequently, as I broaden this discussion to encompass more than remote computing, I'll take extra care to make it clear which computer involved in an on-line session I'm referring to when I use the term *remote system* or *remote computer*.

I'll assume, here, that you do own a modem. As Fig. 10-2 shows, pcANYWHERE initially sets the Modem field in the Quick Connect Parameters dialog box to None. Let's go ahead and change this. To specify a modem type for your Quick Connect settings, use the arrow keys to highlight Modem and press Enter. Press the down arrow to highlight Select from Modem List and press Enter.

pcANYWHERE will display an option box containing the various brands and models of modems it supports, as shown in Fig. 10-3. You can use the arrow keys to page through this list and find your modem type. If the specific model you own

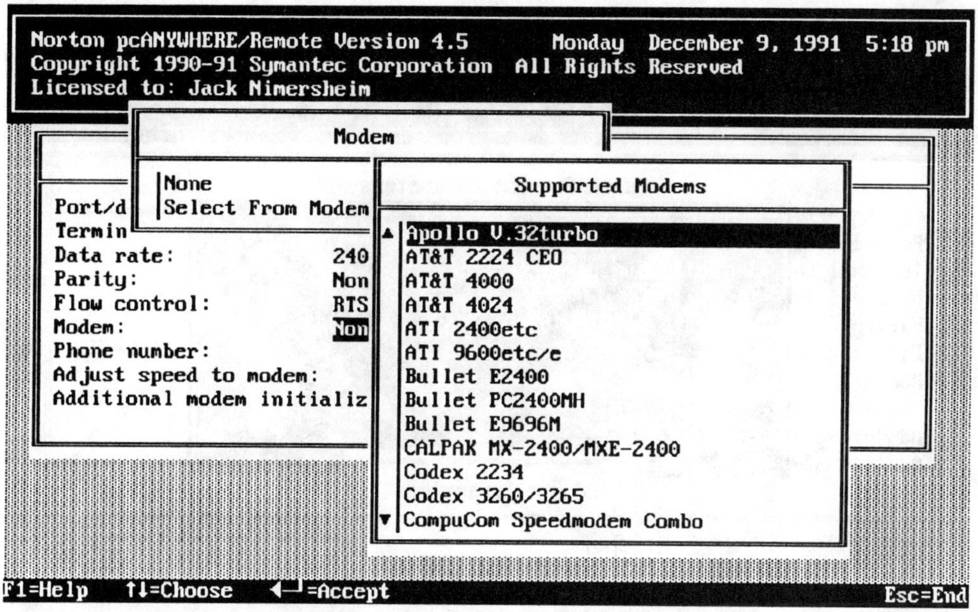

10-3 pcANYWHERE supports a wide range of popular modems.

isn't listed, try choosing a modem it emulates. In most cases, this involves selecting one of the included Hayes' models. (Standards established by this company have been adopted by the vast majority of PC modem manufacturers.) pcANYWHERE's Supported Modem option box includes a Hayes Compatible option, a logical choice if no other selection seems appropriate for your modem.

After you locate and highlight your modem type or an acceptable substitute, press Enter. Depending on the type of modem you select, pcANYWHERE might suggest optimum values for one or more fields in the Quick Connect Parameters dialog box. If it does, you'll be asked if you want to use these values for your default Quick Connect environment. Figure 10-4 shows the settings pcANYWHERE recommends for my modem, a CompuCom Speedmodem Combo. Answering Yes to this prompt causes pcANYWHERE to modify automatically the settings for all listed fields. To set up a generic configuration, you should refuse pcANYWHERE's offer of help.

To ignore pcANYWHERE's suggestions and manually adjust your default Quick Connect environment, use the arrow keys to highlight No and press Enter. If your modem is not connected to COM1, the default setting, you'll also need to let pcANYWHERE know which serial port you're using. To do this:

1. Use the arrow keys to highlight Port/Device Type and press Enter.
2. At Comm Device, press Enter to accept Serial.
3. At Serial Ports, highlight the right COM port for your modem. Press Enter.

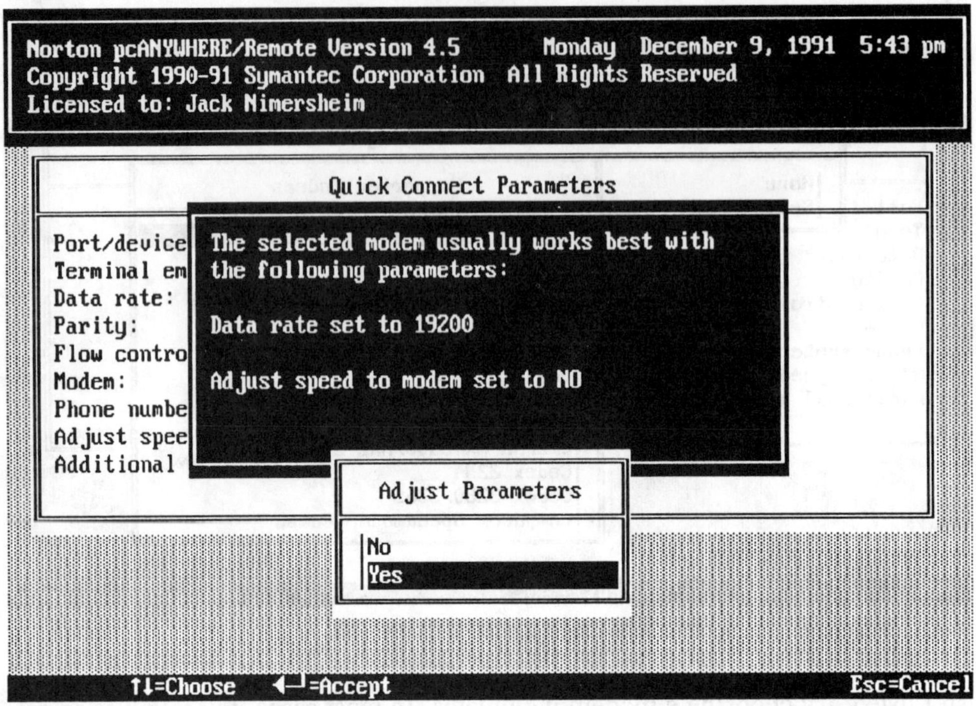

10-4 When appropriate, pcANYWHERE suggests the most efficient parameter settings for a selected modem.

A second assumption I'm going to make here is that you won't be using Quick Connect for pcANYWHERE/Host-to-pcANYWHERE/Remote sessions. Instead, as I suggested earlier, I'm going to work from the premise that you're setting up the current parameters to quickly contact more traditional on-line services, which might include:

- commercial information services such as CompuServe, GEnie, Delphi, and the like
- local bulletin board systems (BBSs)
- dedicated electronic messaging services such as MCI Mail and AT&T Mail
- on-line research services such as Dialog, Lexis, the Dow/Jones News Retrieval, and so forth

Because these services won't be running pcANYWHERE/Host, you'll need to change the Terminal Emulation field to something other than its default setting, pcANYWHERE. To choose a different terminal emulation, use the arrow keys to highlight Terminal Emulation and press Enter.

Selecting this option causes pcANYWHERE to display an option box containing the various types of terminals it's capable of emulating, as shown in Fig.

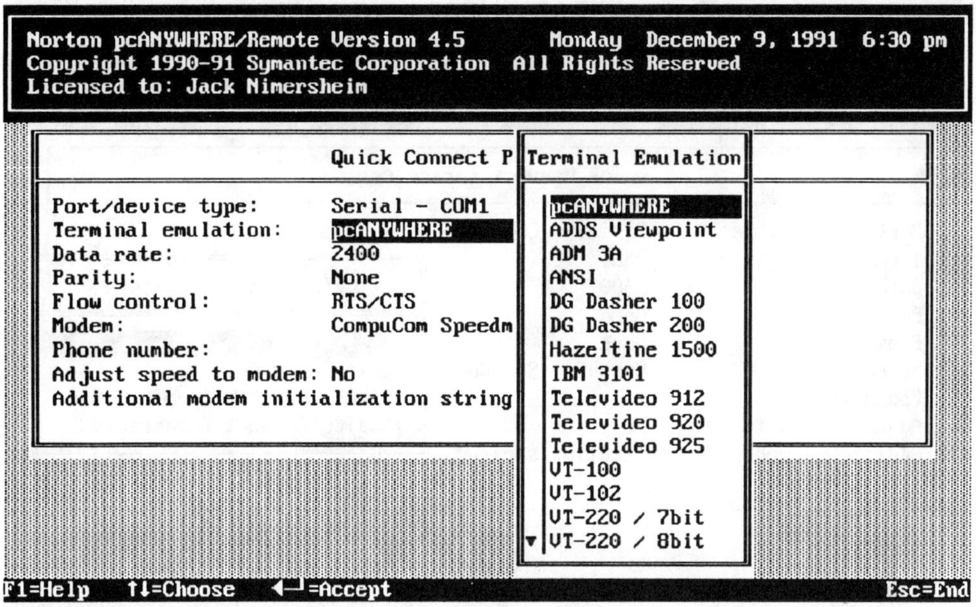

10-5 pcANYWHERE can emulate a number of different terminal types.

10-5. Many of the options listed here are standard terminals used with popular mainframes or minicomputers, and might be appropriate for a specific entry in your Host List. Keep in mind, though, that your primary goal here is to set up a "pure vanilla" environment. The most generic terminal type in this list is the ANSI option.

To select ANSI terminal emulation, use the arrow keys to highlight ANSI and press Enter. Finally, change the Flow Control setting to None, the appropriate setting for most modems operating at speeds slower than 9600 bps. To modify the Flow Control setting, use the arrow keys to highlight Flow Control and press Enter. Use the arrow keys to highlight None and press Enter. At this point, you're ready to modify Quick Connect's initial configuration to incorporate your new settings. To do this, press Esc.

pcANYWHERE will store the new settings as your default Quick Connect configuration and return to its Options menu, as illustrated in Fig. 10-6. But wait a minute! This is a different menu than the one you saw earlier, in Fig. 10-1. Specifically, it contains a new option, Call.

There's really no mystery here. pcANYWHERE includes the Call option whenever a modem is specified in the default Quick Connect configuration. The initial Modem setting, you might recall, was None, a setting that made this option somewhat superfluous because it's impossible to place outgoing calls through a direct connection. Now that you have this capability, however, pcANYWHERE went ahead and added Call to your Quick Connect Options menu. (Considerate little program, isn't it?)

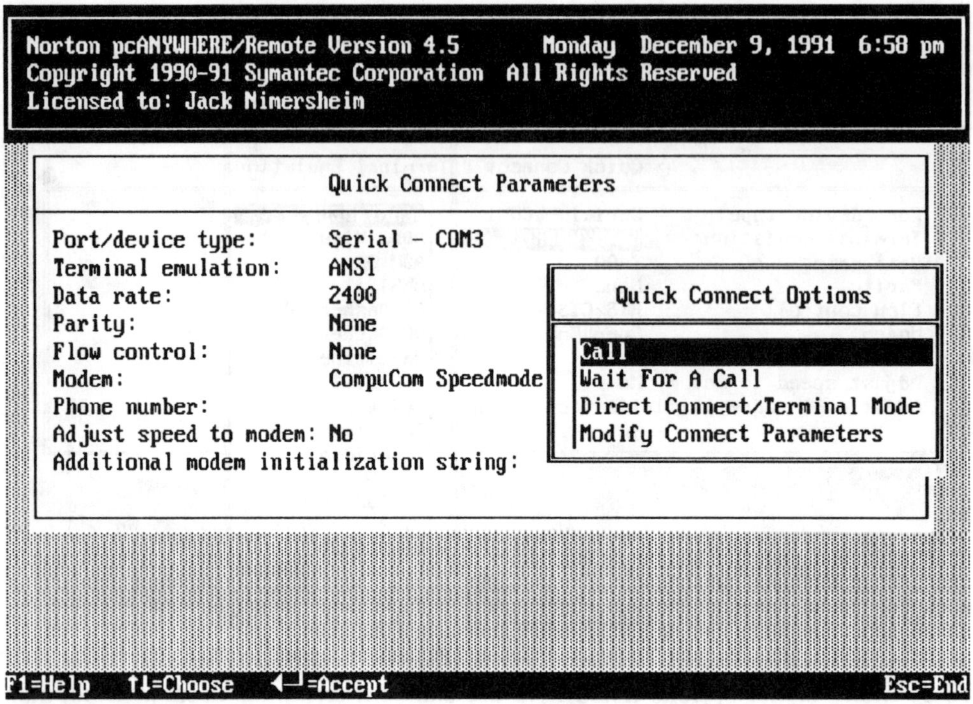

10-6 The options available on the Quick Connect menu depend on your parameter settings.

Calling a remote system

Now that you have this Call option available, why not go ahead and use it? Why not place that first call to a non-pcANYWHERE host so you can see how pcANYWHERE handles more traditional on-line activities?

But which system?

One recurring problem an author faces when putting together a book on telecommunications is how to provide concrete, interactive examples of what the reader can expect to see once he finally has everything up and running and is ready to call another computer with his new setup. Theoretical discussions and objective descriptions can accomplish only so much. If you really want to see what telecommunications is all about, nothing helps as much as empirical, hands-on experience. And the best way to obtain this is by actually contacting a legitimate on-line service and seeing, first hand, what happens.

But which system can I have you call? Commercial on-line services like CompuServe, Delphi, and GEnie restrict access to paid subscribers. (Don't believe for a minute, however, that I plan to ignore these services. Later sections of this chapter will provide concrete examples of what they offer and how they work. Additionally, you'll find information on how to subscribe to some of the more popular commercial on-line services in appendix A, *What's out there?*) A local bulletin board system (BBS), on the other hand, though generally open to the public, simply isn't set up to handle the number of calls that could result from its inclusion in a book that might (hopefully!) be bought by the thousands of individuals who could potentially contact that BBS when performing the hands-on tasks it contains.

I'm intimately familiar with this conundrum, this being the seventh book I've written that includes information on telecommunications. I'm also lucky in that, due to contacts I've developed in my secret identity of a mild-mannered computer journalist, I've also been able to pull a few strings and come up with what I hope is a workable solution to this problem. Before continuing, therefore, I'd like to thank Vulcan Publications and Doug Kilarski, the Editor-in-Chief of *Computer Monthly* (a magazine that actually pays me to write about telecommunications) for agreeing to let me use their on-line service, the Computer Monthly Electronic Edition, or CMEE, for the following exercises. Because of these folks, I can provide you with specific examples of the steps commonly involved in using your modem to access a remote system. Thanks, gang.

Note: One thing that's beyond my capabilities, however, is convincing telephone companies like AT&T or Sprint to suspend any long-distance charges you'll incur when calling distant systems. Consequently, be prepared to see a small increase in your phone bill for the time you spend on-line performing the exercises in this book that require a long-distance call, as does contacting CMEE. Don't panic, though. Wherever possible, I've structured your on-line activities to keep this time to a minimum. Still, long-distance charges and access fees are an unavoidable fact of life for anyone who telecommunicates with any regularity, so you might as well start getting used to them now.

CMEE provides a variety of on-line services, including a hardware and software buyer's guide; a convenient way for readers of *Computer Monthly* to submit suggestions, comments, questions, and the like to the magazine's editors and writers; and special areas for downloading and uploading files to and from the CMEE system. Before accessing any of these features, however, you'll need to use your modem to call CMEE and register with the system. As is true of most on-line services, this involves providing some basic information about yourself and then defining the procedures CMEE should use to recognize you in the future.

Getting ready

It's time to start applying some of the technical information I've been discussing up to this point, so stay with me on this one. To begin with, CMEE uses the following communication parameters:

Baud:	1200, 2400, or 9600 bps
Parity:	None
Terminal emulation:	ANSI

Before calling CMEE, therefore, verify that these are the settings in your current Quick Connect configuration.

As the above listing illustrates, CMEE will handle modem calls placed at either 1200 or 2400 bps, the two most popular speeds currently used for PC communications. (Additionally, a 1200-bps access line can accommodate 9600-bps connections.) Which one you should select depends on the maximum baud rate your modem supports.

Placing the call

Once pcANYWHERE is configured properly, you're ready to make that initial registration call. For this exercise, use the Quick Connect Call option to access the appropriate CMEE number for your modem's baud rate. The specific number you'll enter depends on which baud rate you select.

To initiate this first call, use the arrow keys to highlight Call and press Enter. If you had entered a telephone number in the Phone Number field of the Quick Connect Parameters dialog box, pcANYWHERE would have dialed it automatically. Keep in mind, though, that I was helping you set up a generic telecommunications environment, so you didn't. Because pcANYWHERE doesn't know what number you want to call, it will prompt you to specify one, as shown in Fig. 10-7. Use this prompt to enter the appropriate CMEE number for your modem speed.

Note: If you live in or around Birmingham, Alabama, the corporate headquarters of Vulcan Publications, Inc., contacting CMEE might be a local call. If

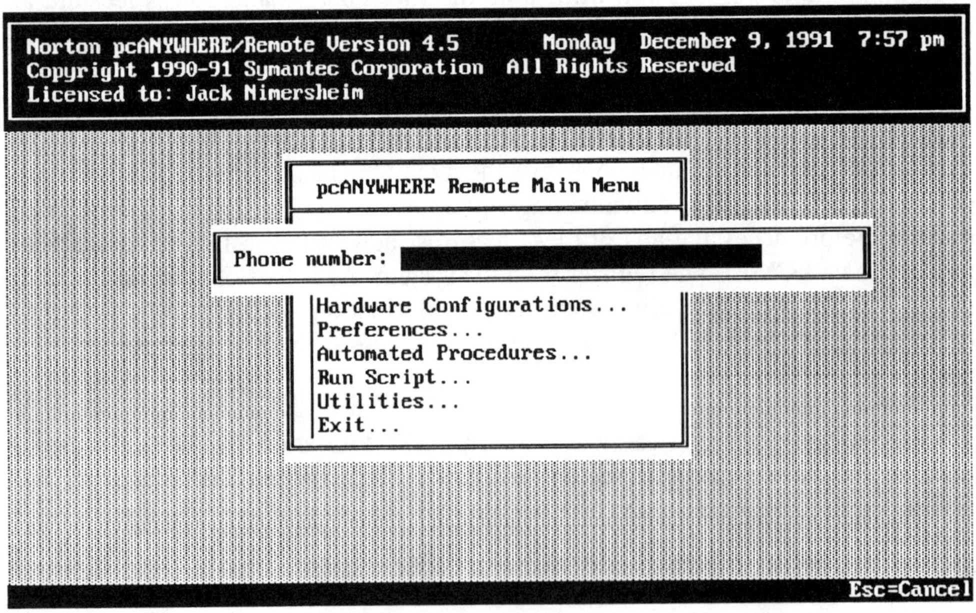

10-7 pcANYWHERE requests the phone number you want to call.

you don't normally include an area code when dialing that city, therefore, feel free to delete the 1-205- portion of the following exercises.

If you own a 1200- or 9600-bps modem, type 1-205-655-4065 and press Enter. If you own a 2400-bps modem, type 1-205-655-4059 and press Enter. Your modem will then begin dialing the number you specified to contact CMEE at the appropriate baud rate. Once pcANYWHERE begins dialing, you'll see the Connection Status screen shown in Fig. 10-8. This screen provides ongoing information about the specific actions that pcANYWHERE must perform to complete the requested call. If the line you dialed is available—that is, if pcANYWHERE doesn't encounter a busy signal—CMEE will automatically accept your call and, after a few seconds, display the welcome message shown in Fig. 10-9.

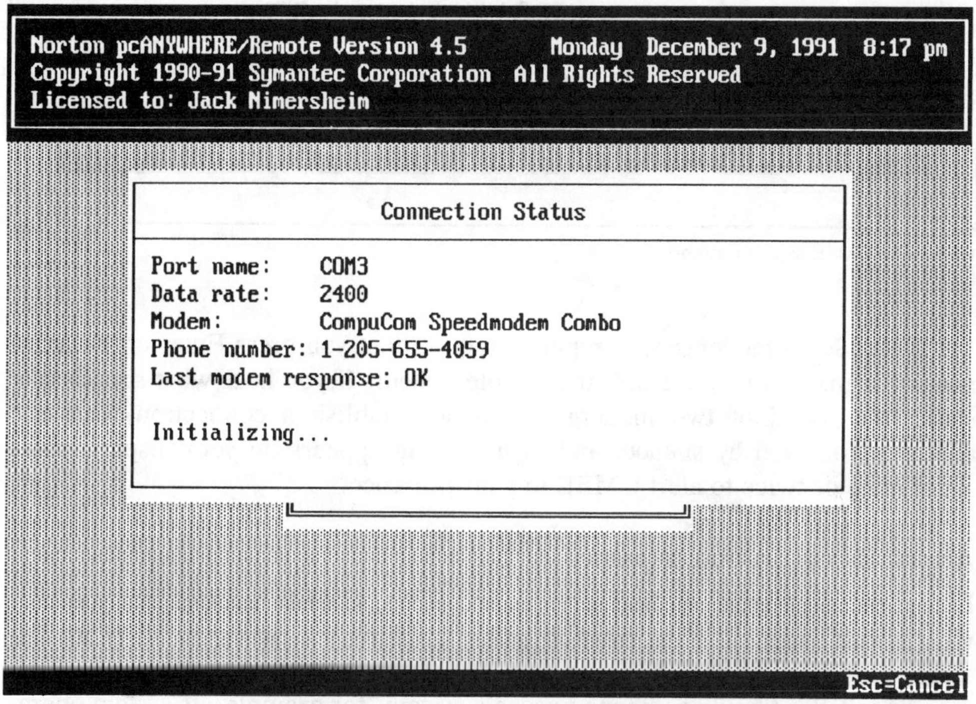

10-8 As it dials, pcANYWHERE displays the Connection Status message.

Pause for a moment and ponder what you just accomplished. Although it appears on your display, this message is being sent to pcANYWHERE over a phone line from CMEE, an electronic bulletin board in the beautiful Southern city of Birmingham, Alabama. Congratulations! You just made your first successful on-line connection to a non-pcANYWHERE system. That wasn't too difficult, was it? Now that you've established this initial connection to CMEE, why not go ahead and register with the system?

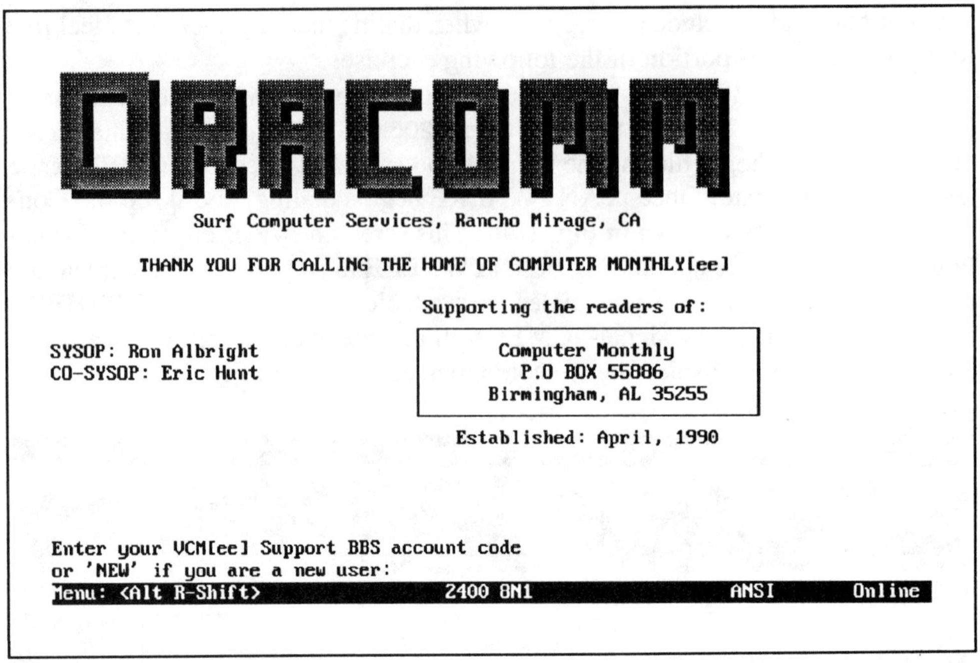

10-9 The CMEE sign-on screen.

Note: Some modems or computers require that you press Enter once a connection is made, to "wake up" the remote system. If you hear what sounds like static (the sound of two modems trying to establish a connection with one another), followed by silence, and then nothing appears on your display, press Enter once or twice to alert CMEE to your presence.

A typical registration sequence

Like most commercial on-line services and BBSs, CMEE requires that you provide it with some basic information before granting you access to all of its features. This is done both for your own protection and for the protection of the service or BBS. Should someone misuse a system, for example, its system operator (or sysop) would be able to identify that individual and take the appropriate steps to prevent him from doing so again. Conversely, registering for a system allows you to specify an account code (also called a user ID) that, in conjunction with a private password you select, allows a system to recognize you as a valid user whenever you contact it in the future.

Although the actual procedures involved in this differ from system to system, registering for CMEE will give you a good idea of what to expect when you place your initial call to different on-line services. First, though, you'll need to get past the CMEE welcome screen.

Registered members of CMEE use the screen in Fig. 10-9 to enter their account code, or *user ID* as it's more commonly called. Being a first-time caller, however, you'll need to complete your registration before being able to do so.

To register as a new user to CMEE, type NEW and press Enter. Identifying yourself as a new user will advance you to the prompt shown in Fig. 10-10. This prompt asks you to enter an account code, the first piece of information CMEE requires to process your registration.

```
Enter your VCM[ee] Support BBS account code
or 'NEW' if you are a new user: new

To setup an account, you will be asked to enter 8 characters
(letters or numbers) to be used as your account code.

CHOOSE THIS CAREFULLY, as this account code is the way by which
you will be known to other users, and by which they will address
mail to you.  If you cannot think of an appropriate eight
characters to be used as your account, you might want to hang up,
give it some thought, and call back.

   DO NOT PUT YOUR TELEPHONE NUMBER
         IN YOUR DESCRIPTION
       OR IN A PUBLIC MESSAGE!!

REMEMBER YOUR ACCOUNT AND PASSWORD.  If you forget your account or
password, you will not be able to access the system again!
Accounts are deleted if you do not log on again within two weeks.


Enter 4 to 8 letters to be used as your account code:
Menu: <Alt R-Shift>               2400 8N1                 ANSI       Online
```

10-10 Your account code (or user ID) identifies you to an on-line system.

Your account code—or, as I pointed out above, user ID—is the unique name you want assigned to your account. One common approach many people use when selecting an account code is to select some variation on their name. In the exercises that follow, for example, I'll use JACKN as my account code. To specify your CMEE account code, type in a user ID, four to eight letters in length, and press Enter.

At this point, CMEE will display a prompt asking you to input a password that will be assigned to your account code. As is true in pcANYWHERE, passwords you select for other on-line services guard against unauthorized callers accessing your account on a particular system. It's wise to take full advantage of this security device, given that most systems charge you for their services. The best approach is to select a password that isn't obvious, but can be easily remembered. I don't want to instill paranoia here, but anyone who knows your User ID and password for a specific system will be able to access your account on that system, appearing for all intents and purposes like you whenever they sign on. That's why I (and most other so-called experts) recommend using an obscure password. Avoid including a variation of your name, hometown, spouse's name, or anything else that would be easy for a resourceful person to figure out, in your passwords. Also, it's a good idea to select a different password on each remote system for which you're registered—once again, for security's sake.

To specify your CMEE password, type an appropriate password (four to eight letters in length) and press Enter. After recording your account code and password, CMEE will request additional information from you, such as your name, the city and state in which you live, and a short, three-line description you'd like associated with your account. The prompts associated with each of these items are self-explanatory, so go ahead and fill in this information now. To complete your CMEE registration:

1. Type in your full name and press Enter.
2. Type in the city and state in which you live and press Enter.
3. Enter a three-line description of yourself, pressing Enter at the end of each line.

Note: Most on-line services provide some means for callers to "get to know" other registered users. CMEE will display the information you're entering here whenever another user requests information about you, based on your account code. Shortly, CMEE will ask a few more questions about you. This additional information is never made public and can be accessed only by the system operator. This approach, where certain information is made available to other callers while other information is collected for internal use only, is fairly common among on-line services.

Once you've entered your preliminary information, CMEE will request that you verify the account code and password you specified by typing it in a second time. Although the system displays your account code as you enter it, you'll notice that it will substitute a series of asterisks (*) for your password, as shown in Fig. 10-11. This is done for security purposes, so that someone else doesn't see your password during the normal sign-on procedure.

You're now officially signed onto CMEE and should see a screen resembling Fig. 10-11. The specific announcements on the CMEE opening screen are regularly updated by system operator Dr. Ron Albright, and will differ from the ones shown here. To continue the current CMEE session, press Enter.

CMEE is a multiline system, meaning that more than one person can be working with it at any given time. Each time you call CMEE, one of the first things it does is display a listing of any users currently on-line, including yourself, as shown in Fig. 10-12. (Like most on-line services, CMEE includes a chat fea-

```
Enter your account code: jackn
Enter your password: ********

Good evening, Jack Nimersheim.    Welcome to VCM[ee] Support BBS.


NEW! Two new communications programs for the IBM: GO and PILOT!
     OUTSTANDING Shareware that makes modem communications EASY!
++++++++++++++++++++++++++++++++++++++++++++++++++++++++++++++++++++++++
NEW!! GALAXY! Commander Keen's sequel from Apogee...the GREATEST YET!
NEW! SHADES 3-D space adventure for EGA/VGA. EXCELLENT Shareware!
NEW! ALDO3.ZIP - best arcade game out for the files size! AWESOME!

        These files and more are found in the NEW DOWNLOADS section.
        Type "10" at any "What next?" prompt and "FB" to view list.
++++++++++++++++++++++++++++++++++++++++++++++++++++++++++++++++++++++++

Press [ENTER] to continue; N for nonstop;
Menu: <Alt R-Shift>                  2400 8N1                 ANSI       Online
```

10-11 You must enter your ID and the correct password to access most on-line systems.

```
NEW! Two new communications programs for the IBM: GO and PILOT!
     OUTSTANDING Shareware that makes modem communications EASY!
++++++++++++++++++++++++++++++++++++++++++++++++++++++++++++++++++++++++
NEW!! GALAXY! Commander Keen's sequel from Apogee...the GREATEST YET!
NEW! SHADES 3-D space adventure for EGA/VGA. EXCELLENT Shareware!
NEW! ALDO3.ZIP - best arcade game out for the files size! AWESOME!

        These files and more are found in the NEW DOWNLOADS section.
        Type "10" at any "What next?" prompt and "FB" to view list.
++++++++++++++++++++++++++++++++++++++++++++++++++++++++++++++++++++++++

Press [ENTER] to continue; N for nonstop;

You can be online 60 minutes this session.


Users Currently Online
  Line Account  Cmnd
    1  HENRY     FF
    2  JACK4     SM

Do you want a demonstration of how to use this BBS ([Y]/N)?
Menu: <Alt R-Shift>                  2400 8N1                 ANSI       Online
```

10-12 CMEE is a multiline BBS.

ture, which allows you to communicate directly with any other callers included in this list. Chatting with your fellow callers is a great way to make on-line friends.) Because this is your first time on the system, CMEE also asks whether you're interested in seeing a demonstration of its bulletin board software, the Oracomm BBS.

If you feel like exploring the Oracomm demonstration, type Y and press Enter. Note: Keep in mind, however, that this will extend the time you spend on-line and, by extension, your long-distance bill for this initial session.

To bypass this demonstration, type N and press Enter. Because you're a first-time user, CMEE will request additional, private information before advancing you to its main services, as shown in Fig. 10-13. As I mentioned earlier, the information you enter during this sequence is not posted for public use and can be accessed only by authorized system managers—specifically, sysop Dr. Albright and his colleague, Eric Hunt, the two gentlemen responsible for managing the CMEE bulletin board. Information you enter here includes:

- your real name
- your area code and telephone number
- your street address
- your city, state, and zip code

That's the full extent of the personal information CMEE requests, certainly no more than any legitimate caller should feel comfortable supplying. As the mes-

```
Press [ENTER] to continue; N for nonstop;

You can be online 59 minutes this session.


Users Currently Online
  Line Account  Cmnd
    2  JACK2     SM

Do you want a demonstration of how to use this BBS ([Y]/N)? n

You must answer a questionnaire before proceeding.

Your answers may span more than one line.
Press [ENTER] on a blank line when you are finished
answering each question to go on to the next one.

This is a free BBS, but , to have access to all areas,
*you must answer some questions truthfully.  This information will not be
*made available to other users or anyone else.  It is used to be verify
*who is on the system to prevent illegal information from being posted.
*What is your real name:

Menu: <Alt R-Shift>           2400 8N1            ANSI        Online
```

10-13 Most services request additional information, which is available only to the system operators.

sage in Fig. 10-13 indicates, this information is required strictly to verify who is on CMEE and to provide the sysops with a way to identify any callers who might misuse the system.

After entering your personal information, you'll see the part of CMEE's main menu, shown in Fig. 10-14. To see the rest of this opening menu, press Enter.

```
What is your mailing city, state, zipcode:
Vevay, IN  47043

   HENRY: Just logged on.  2 lines in use.
   JACK3: Just logged on.  2 lines in use.


Messages TO you:
No Messages

Press [ENTER]


*** VCM[ee] Support BBS Main Command Menu ***
R - Read a message
E - Enter a message
W - Who are the other users
B - subBoard Selection Menu
T - Tutorial: How to use this BBS
[ENTER] - for remainder of the menu


12/09 20:29 1/59min  Personal Mail Board (#1)
What next?
Menu: <Alt R-Shift>              2400 8N1              ANSI      Online
```

10-14 Congratulations! Your sign-on was successful.

Ending your first session

Again, congratulations! You're now a registered user of CMEE and can call back at any time, without having to reenter all of the information supplied in the previous exercises. From now on, after supplying your account code and password, you'll be taken almost immediately to the system's Main Command menu, shown in Fig. 10-15. Once this menu is displayed, you'll be able to select from a wide range of options representing the various on-line activities CMEE supports. Now that you're registered, you can sign on to CMEE at any time and explore these various offerings. For now, however, use the Goodbye option to end this initial CMEE session, go off-line, and return to pcANYWHERE.

To end your CMEE session, type G and press Enter. When the system asks if you want to leave a message for the sysop, type N and press Enter. So you've

```
W - Who are the other users
B - subBoard Selection Menu
T - Tutorial: How to use this BBS
[ENTER] - for remainder of the menu


12/09 20:29 1/59min  Personal Mail Board (#1)
What next?


S - Summary of messages
A - Space Gamble!
N - Change name/description/password
U - Upload a file to the bbs
D - Download a file from the bbs
F - List FILES that can be downloaded
V - View a text file
C - Chat with other users or sysop
H - Display HELPful instructions
G - Goodbye


12/09 20:30 1/59min  Personal Mail Board (#1)
What next?
Menu: <Alt R-Shift>              2400 8N1                 ANSI        Online
```

10-15 The CMEE Command menu.

signed on and then signed right back off again. Seems kind of unproductive, doesn't it? Not really. You actually accomplished quite a bit on this initial call:

- First, you learned how to contact a remote system with pcANYWHERE's Quick Connect option.
- Second, you got a general idea of the procedures used to sign on to a remote system.
- Third, you managed to register yourself as an authorized user with CMEE. Consequently, whenever you place another call to this particular system and enter your User ID and password, you'll be taken almost immediately to its main menu.
- Finally, you learned how to end an on-line session with the CMEE GOODBYE command.

Not bad for a few minutes of your time and a few pennies in long-distance telephone charges, I'd say. Furthermore, although the commands used to accomplish specific activities will vary from system to system, the general procedures outlined in the previous exercises are common to virtually all commercial on-line services and BBSs.

Perhaps most important, however, is the fact that you've now dipped your toes into the pool of telecommunications. In the process, it should have become clear to you that this isn't really some mysterious or mystical experience. You're simply using several modern tools—your PC, a modem, and pcANYWHERE—to accomplish some very impressive feats. And you're not finished yet. In the next section, you're going to submerge yourself even deeper into the waters of telecommunications, both figuratively and technically, as you drop one level further down and use another Quick Connect option to communicate directly with your modem, a process that involves setting up pcANYWHERE to run in terminal mode.

Running pcANYWHERE/ Remote in terminal mode

Signing off the CMEE BBS at the end of the previous section automatically returned you to the pcANYWHERE/Remote Main Menu. In order to demonstrate how terminal mode works, you'll first need to get back to the Quick Connect Options menu.

To display the Quick Connect Options menu, use the arrow keys to highlight Quick Connect and press Enter. Once again, pcANYWHERE/Remote will display the screen shown in Fig. 10-6, which contains the Quick Connect Options menu.

Selecting terminal mode

You've probably already figured out which option switches pcANYWHERE over to terminal mode. To run pcANYWHERE in terminal mode, use the arrow keys to highlight Direct Connect/Terminal Mode and press Enter.

At this point, your screen should resemble Fig. 10-16. If this screen looks familiar, it's because it's exactly what you saw when you signed off CMEE at the end of the previous exercise. Explaining why this is so will provide useful information about how terminal mode works and what it allows you to do.

What is terminal mode?

You might not realize it, but you've already used pcANYWHERE in terminal mode. Consider what telecommunications is all about: Using your PC, a modem and a program like pcANYWHERE to contact another computer. For this to happen, your PC and modem must be able to communicate with one another. That is, information from your PC must be passed on to your modem so that, in turn, it can be transmitted across the phone line to the other system.

Most of the time, pcANYWHERE handles this automatically—as it did, for example, when it issued the various commands required to dial CMEE after you selected the Quick Connect Call option. Once this connection was established, however, pcANYWHERE temporarily bowed out of the picture and turned control of your modem directly over to you. From that point on, anything you typed in at your keyboard was transferred immediately to your modem for subsequent transmission to CMEE.

Selecting Direct Connect/Terminal Mode restores this direct communication between you and your modem—although, in truth, it was never terminated. After signing off CMEE, pcANYWHERE merely temporarily reinserts itself between you and your modem so it can ask, through its menus, what you want to do next. Because you're choosing to return to terminal mode immediately, the final screen from your previous connection survives intact. So where are you now? Right back where you were before, in direct communication with your modem.

Note: If you're concerned about long-distance charges, don't be. Although it might appear as if you're still hooked into Birmingham, this is one instance where looks definitely are deceiving. What you're seeing in Fig. 10-16 is the last screen CMEE transmitted to your PC before "hanging up." In fact, the NO CARRIER message at the bottom of this figure was generated by your modem when it recognized that a connection no longer existed. So don't worry. The bean counters at AT&T (or Sprint, or whichever long-distance carrier you use) no longer have their hands in your pocket.

The next logical question becomes, "Now that I'm in direct communication with your modem, what can I do?" Stated simply, you can do manually almost everything that, up until now, pcANYWHERE has done for you automatically—assuming, of course, that you know the appropriate commands.

I'll get to those commands in a few paragraphs. First, however, I want to

```
12/10 09:57 0/60min  Personal Mail Board (#1)
What next? g


Do you want to leave a comment for the SYSOP (Y/[N])? n

Copyright 1988-1991, Surf Computer Services  - All Rights Reserved
Surf Computer Services, 71-540 Gardess Rd, Rancho Mirage, CA 92270
To run your own Oracomm BBS, write to the above address.


      TRY THESE OTHER terrific BBSs - Members of:

           Birmingham Association of BBSs (BAB)

                   The ST-BBS    --  205-836-9311

           Disktop Publishing    --  205-854-1660

      You'll be glad you dropped in on these fine & fun systems!!
räPT|ºm½÷µ
NO CARRIER

Menu: <Alt R-Shift>                 2400 8N1               ANSI         Offline
```

10-16 A NO CARRIER message indicates that the phone connection has been terminated.

remove the clutter from your display and, in the process, introduce another pcANYWHERE feature.

The pcANYWHERE Terminal Options menu

Remember in the previous chapter, when you used the Session Options menu to coordinate activities between your host and remote systems? pcANYWHERE provides a similar menu for simplifying communications between you and your modem whenever you're operating in terminal mode. To display the Terminal Options menu, press Ctrl−Right Shift, the Menu hot key.

Tip: Keep in mind that, during installation, pcANYWHERE defines Ctrl-Right Shift as the default hot-key command used to display its on-line menus. If you've used pcANYWHERE/Remote's General Preferences option to select a different default hot-key combination, you'll need to substitute that command in the above exercise to display your Terminal Options menu. (You'll notice, for example, that I redefined my Menu hot key to be Alt−Right Shift, which I did to avoid conflicts between pcANYWHERE and the screen-capture program I'm using to generate the figures in this book.)

Pressing the Menu hot key with pcANYWHERE running in terminal mode displays the Terminal Options menu, shown in Fig. 10-17. Options available from this menu include:

Break Selecting this option transmits a break signal to the other system involved in an on-line session. What this does depends largely on the type of system you're connected to and what you're doing at the time. Many mainframe operating systems, for example, use a break signal to interrupt the current operation, much like DOS uses a Ctrl−C.

End Session This option works identically to its counterpart in the Sessions menu, in that it terminates your connection with the other system.

File Transfer This is another option that duplicates its Session menu counterpart. You use it to initiate a file transfer between the two computers involved in an on-line session.

Go to DOS Again, a duplicate item. This option temporarily suspends terminal mode and opens pcANYWHERE's DOS shell.

Monitor On This option toggles a special display mode on and off. When it's on, signals coming across a modem link that normally are not displayed on your monitor—line feeds, carriage returns, and the like—will be translated into visible characters, a useful tool for analyzing data transmitted by the other computer. Selecting this option a second time toggles off display translation.

Print On Use this option to toggle your printer on and off to capture portions of an on-line session.

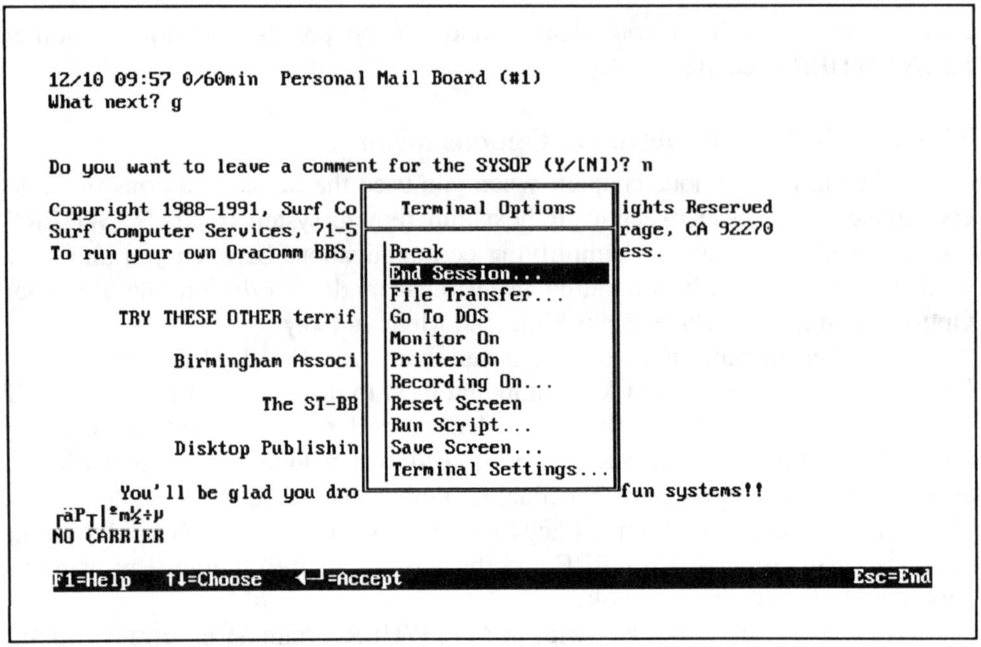

10-17 pcANYWHERE's Terminal Options menu.

Recording On As it did in the Session menu, this option toggles session recording on and off.

Reset Screen That's pcANYWHERE's fancy way of saying "clear screen," folks. Selecting this option clears your display and returns the cursor to the top-left corner of the screen. (Selecting this option is like issuing a DOS CLS command.)

Run Script Yes, you can automate on-line activities performed in a traditional on-line service. This option makes it possible. (I'll remind you once again that I'll be covering pcANYWHERE's script language in Part 3 of this book, *Beyond the basics*.)

Save Screen This option lets you save a "snapshot" of the current contents of the remote display to a disk file for future reference. As you'll see in chapter 13, pcANYWHERE includes a special utility that can be used to recall saved screens.

Terminal Settings Selecting this option displays a dialog box containing several settings that influence how the two computers involved in an on-line session interact with one another. Items in the Terminal Settings dialog box include COM port settings, terminal emulation, data rate (bps setting), parity, and the like. Be aware, however, that modifying these settings during an on-line session can result in the connection being terminated.

Most of the Terminal Options are designed to be used during an actual on-line session—that is, when you're using pcANYWHERE's terminal mode to communicate with another computer. Some, however, also can be useful when it's just you and your modem "talking" to one another. For example, there's no real need to keep that final message from your previous CMEE session on your display. It's only cluttering things up. So get rid of it.

To clear your display, use the arrow keys to highlight Reset Screen and press Enter. At the subsequent Reset prompt, press Enter to indicate Yes. This will clear the display and return the cursor to the upper left corner of your screen. Now that you have a true "clean slate" to work with, let's get to work figuring out how to communicate directly with your modem when running pcANYWHERE in terminal mode.

Using AT commands

As mentioned earlier in this chapter, most PC modems are designed to be Hayes-compatible—that is, they adhere to standards established years ago by Hayes Microcomputer Products, Inc., a popular modem manufacturer. Among the most critical tests of whether or not a modem can be considered Hayes-compatible is how well it supports the so-called AT command set. Due to the popularity of the Hayes modem line, its internal command structure, the AT command set, quickly became the standard for PC communications. Today, virtually every modem manufacturer strives to make their products Hayes-compatible, meaning they recognize and respond to the AT command set.

What are AT commands? The AT command set derives its name from the fact that you precede instructions sent to a modem from your PC with the letters AT, a special code designed to get the modem's *at*tention. For example, the AT command used to tell a modem to dial a telephone number using the touch-tone method is ATDT (*at*tention *d*ial *t*one). When a modem is being used with a phone system that doesn't support touch-tone dialing, the corresponding AT command is ATDP (*at*tention *d*ial *p*ulse). Like these two examples, many AT commands are mnemonic in nature, meaning the command itself reflects the operation being performed. Other mnemonic AT commands include:

ATA (attention answer) Forces a modem to answer an incoming call, even if it's not set to auto-answer mode.

ATB (attention bell protocol) Specifies handshake options, using standard Bell protocols.

ATE (attention echo) Turns on and off the local echo of characters during an on-line session.

ATH (attention hangup) Places the modem on and off the hook—that is, connects or disconnects a modem from the phone line.

ATO (attention return on-line) Returns a modem to on-line status following command execution.

If all AT commands were this obvious (and, therefore, this easy to memorize), issuing commands to your modem while running pcANYWHERE in terminal mode would be a pretty straightforward proposition, indeed. You'd issue an easily remembered command, your modem would respond, and that would be the end of it. Unfortunately, this isn't the case. To prove my point, let's take a look at some other AT commands that aren't so obvious:

ATZ Reconfigures a modem to its default settings.

ATV Specifies whether a modem should display the results of specific operations as a numeric code or text message.

ATM Specifies when a modem's speaker is active; common choices are Never, Only Until a Connection Is Established, or Always.

ATL Internally sets a modem's speaker volume; some modems have an external knob, similar to the volume control on a tape player, that serves this same function.

Hardly intuitive, eh? Then there's the AT Set Register command (ATS), a single command that can modify several modem settings, depending on the value that follows it. For example:

ATS0 Specifies how many times the phone should ring before your modem answers the line when set to auto-answer mode.

ATS8 Specifies how long your modem should pause whenever it encounters a comma in a dial string.

ATS9 Sets how long your modem should wait to terminate a call if it doesn't reach a modem on the other end.

AT10 Sets how long your modem should wait after the other computer drops its signal before hanging up the line on your end.

As these few examples demonstrate, the Hayes AT command set is a powerful (although admittedly confusing) method of communicating directly with your modem. And I've only scratched the surface here. There are literally dozens of AT commands available. Your modem's manual should contain a complete listing of the ones it recognizes.

Cutting through the clutter

While the AT command set admittedly superimposed a "standard" over the previously chaotic field of telecommunications, using that standard is not an activity to be undertaken by the uninitiated or faint of heart—as the previous examples so aptly illustrate. So, must you become an expert at using the AT command set to run pcANYWHERE in terminal mode? Fortunately, no. As you've probably already realized, the primary function of the various pcANYWHERE Quick Connect and Terminal Options menu items is to insulate you from having to bother with the AT command set.

You don't need to know, for example, that ATDT is the AT command for dialing your phone if your local system uses touch-tone dialing. pcANYWHERE transmits it automatically to your modem when you specify a telephone number after selecting the Quick Connect Call option.

There might be times, however, when you'll be forced to use an AT command. Primarily, this will happen because no corresponding pcANYWHERE menu selection is available. Alternately, in certain situations, using the AT commands set can be more convenient than accomplishing the same thing using the flexible but admittedly complex pcANYWHERE menu structure. Suppose, for example, that you try to contact CMEE using Call, but encounter a busy signal. As convenient as it is, Call still forces you to enter another computer's number each time you try to dial it. In this case, therefore, you would have to keep entering the Birmingham long-distance number until a connection is made, not a very attractive alternative.

What if, instead, there were an AT command that allowed you to automatically redial CMEE's number, on demand, until you got through? Wouldn't that be more convenient? There is, and it is. You'll see how this works in a moment. Before getting to that, however, let's see how the AT command set can be used to dial CMEE's number—or any phone number, for that matter—in the first place.

The "direct" approach

To demonstrate this procedure, I'm going to have you use pcANYWHERE's terminal mode to call . . . well . . . yourself. That might sound a little weird but, as you'll see in a minute, it's the perfect way to demonstrate how the AT commands can be used to redial a number.

To enter the AT dial command on a touch-tone line, type ATDT and press Spacebar. To enter the AT dial command if your telephone system uses pulse dialing, type ATDP and press the spacebar. Next, type in your telephone number in the format *###-####* (where each # represents a digit of your phone number) and press Enter.

Depending on the type of modem you have, and how that modem and your communications program are configured, one or more of the following things should have happened:

- If your modem has a speaker, and that speaker is set to remain on at least until carrier detect—that is, until a connection is established—then you would have heard a dial tone, followed by the sounds of the modem dialing your number. (On a touch-tone line, this last item would have been audible tones; with pulse dialing, you would have heard a series of clicks.)

- If your modem is set to return text messages (remember this from the ATV command mentioned previously), then the word BUSY would have appeared on your screen (after all, you were calling your own phone line), and the modem would have automatically dropped the line.
- If your modem doesn't automatically determine line status, then you would have seen no message and your modem would still be connected, vainly trying to complete the call.

The Hang Up command

Of course, you probably don't want to leave your modem just sitting there, listening to an incessant busy signal. The good news is, you don't have to. The AT Hangup command lets you quickly disconnect your modem from a remote system. To disconnect the current connection, type + + + ATH0 and press Enter.

Note: If for some reason the AT hangup command does not disconnect your call, you can accomplish the same thing by turning off your modem and then turning it back on again. Although this procedure should only be used as a last resort, go ahead and do it now, if necessary.

Congratulations! You just attempted your first connection running pcANYWHERE in terminal mode. Never mind that you only called yourself. And never mind that you reached a busy signal. The important thing is, you did it. But let's talk a little bit about what you did. The steps are as follows:

1. Enter the ATD*x* command, where *x* is either a T or a P, depending on the type of phone system you're using.
2. Press the spacebar to insert a space after the AT dial command.
3. Type in your telephone number.
4. Press Enter to transmit the preceding command sequence to your modem.
5. Use the standard AT escape sequence (+ + +) to temporarily interrupt the current session and alert your modem that another AT command is coming.
6. Use the ATH0 command to disconnect the current call.

The space I had you put between the ATD*x* command and your phone number is optional. Any modem that's fully Hayes-compatible will recognize an AT dial command entered in several formats, including:

ATDT*#######*
ATDT*###-####*
ATDT *#######*
ATDT *###-####*

Similarly, valid formats for long-distance calls include:

ATDT1##########
ATDT1-###-###-####
ATDT1(###)###-####
ATDT 1 ### ### ####
ATDT 1-###-###-####
ATDT 1 (###) ###-####

Basically, your modem ignores any non-numeric characters included in phone numbers following an ATD*x* command—spaces, dashes, brackets, and the like. (One important exception to this rule is the comma, which I'll discuss shortly.) I simply find that entering the numbers I'm dialing in a familiar format such as ###-#### helps organize my normally befuddled mind.

The AT Redial command

Now that you've seen how to use the AT dial and hangup commands, let's return to that imaginary scenario I laid out earlier, where you can use the AT command set to conveniently redial a busy number rather than repeatedly recalling the Call option. To see how this is done, you're going to use the AT redial command to call yourself a second time. "If at first you don't succeed . . ."

To retry the previous call, type A/. The A/ command, another member of the Hayes AT family, lets you automatically redial the last number called. (Of course, the previous chain of events will merely repeat themselves—including, possibly, the need to manually disconnect the call with an ATH0 hangup command or by turning your modem off and then back on again. If this is the case, go ahead and do so now.) As I stated earlier, this two-stroke AT command is surely an easier method for continuing your attempts to reach a system returning a busy signal than repeating the several steps required to use the Quick Connect Call option.

As I stated earlier in this section, those occasions when you have to issue AT commands directly will be rare. Most of the time, there will be a pcANYWHERE menu option available that was designed to accomplish the same thing. Still, it's good to understand something about what the AT commands do and how to use them should the need ever arise. For now, leave terminal mode behind you and return to the pcANYWHERE/Remote Main Menu.

To exit terminal mode, press Ctrl−Right Shift (or the appropriate hot-key command) to display the Terminal Options menu. Use the arrow keys to select End Session and press Enter. That's all there is to it. This wakes pcANYWHERE/Remote up from its temporary slumber, as it were, and returns you to the Main Menu.

An on-line sampler

Now that you know how to use pcANYWHERE to reach out and touch a computer inside your own home, you might well be wondering what lies on the other side of those electronic walls. In all honesty, almost anything. As I pointed out in chapter 1, telecommunications places a wide and diverse range of goods and services literally at your fingertips. While I lack the luxury of enough space to examine all of them, some quick examples of the most common on-line activities should provide adequate evidence of the potential value of pcANYWHERE.

Keep in mind, though, that I have room to hit only the highlights here. I'll begin with what is arguably the most popular reason people use commercial on-line services and local bulletin board systems—electronic messaging.

Electronic messaging

As I pointed out in chapter 1, it's estimated that anywhere from 8 to 15 million people currently rely on electronic messaging, or e-mail, for at least a portion of their regular correspondence. And why not? Exchanging information with a computer can be fast, reliable, convenient, and potentially less expensive than regular mail.

When it comes to convenience, it's hard to beat electronic mail. To begin with, you can compose and transmit messages whenever it's convenient for you, just as you pick up and dial the telephone as your schedule permits. Also, you can "post" your electronic message to an on-line service immediately, regardless of where the person to whom it's being sent might be, or what he's doing at the time.

A second convenience unique to electronic mail is that, unlike a traditional phone call, which requires that two people be available at the same time, the store-and-forward nature of e-mail precludes this necessity. In this area, moreover, the convenience extends to both the sender and recipient of an e-mail message. The person receiving your electronic messages is also able to retrieve it at his convenience. Consequently, neither party's important work is interrupted, but correspondence still flows reliably and with relative speed. To be honest, electronic messaging is so convenient that it's almost addictive. I check my various on-line services at least twice a day. Anyone who communicates with me using e-mail, therefore, can rest assured that there's never more than a twelve-hour delay in my getting their important messages.

On-line options

Commercial information services have jumped on the e-mail bandwagon "faster than fleas on a beagle," to quote one of my father's favorite expressions. Virtually every commercial information service includes electronic messaging in its on-line offerings. CompuServe, Delphi, GEnie, BIX—all offer fairly easy to use electronic messaging.

Besides being available as one of many options offered by commercial on-line services, electronic messaging represents the exclusive product of several dedicated e-mail services. The two chief competitors in this market are familiar names: AT&T and Sprint. These two companies, also, are natural candidates for e-mail providers. After all, each already has in place an international communications network, the primary prerequisite to efficiently disseminating electronic messages on a global scale.

Although the specific procedures employed to pass electronic messages back and forth will vary for each of these services, be it a dedicated e-mail provider or one of the more diverse commercial services, the overall process is pretty much standard fare. You enter the appropriate command for whatever service you're using, identify who the current message should be sent to, optionally enter a short description of its contents, write the message, and then send it merrily on its way.

The next time the intended recipient signs on to that service, whether two minutes or two months later, he'll receive a notice informing him that there's a message waiting. Does all this sound familiar? It should. They don't call this e-mail for nothing. In truth, electronic messaging bears a striking resemblance to the steps normally required to write and deliver a letter using traditional mail.

So do you want to see an example of e-mail in action? I figured you might. Why not look over my shoulder as I use pcANYWHERE to complete a typical e-mail transaction over MCI Mail.

e-mail in action

In the following exercise, I'll be using MCI's 800 number to call that service and see if there are any messages waiting for me. If you're an MCI Mail subscriber, feel free to follow along on your own system.

To sign on to MCI Mail, use the arrow keys to highlight Quick Connect and press Enter. Select Call and press Enter. At the Phone number prompt, type 1-800-456-6245 and press Enter.

As before, pcANYWHERE initiates all the steps required to place the requested call. Consequently, it won't be long before you're connected to MCI Mail's central "mail drop," which is located in Washington, D.C. After this connection is made, press Enter to "wake up" MCI Mail.

This will advance you to the initial MCI Mail sign-on prompt, shown in Fig. 10-18. As is true of virtually all on-line services, MCI Mail requires that you verify who you are before it will grant you access to the system. As is common, this involves entering some kind of identification and a unique password.

To sign on to MCI Mail, type your assigned MCI Mail username and press Enter. At the password prompt, type your MCI Mail password and press Enter. If

```
Pad ID: D7 - Port: 22.
Please enter your user name:
```

10-18 The MCI Mail user name prompt.

your username and password identifies you as a valid MCI Mail subscriber, you'll be advanced to that service's sign-on message, shown in Fig. 10-19. Notice that the message on my screen includes an announcement that I have 1 message in my INBOX, the location where MCI Mail stores e-mail waiting for its subscribers. Before reading this message, I'll scan my INBOX and see what it is.

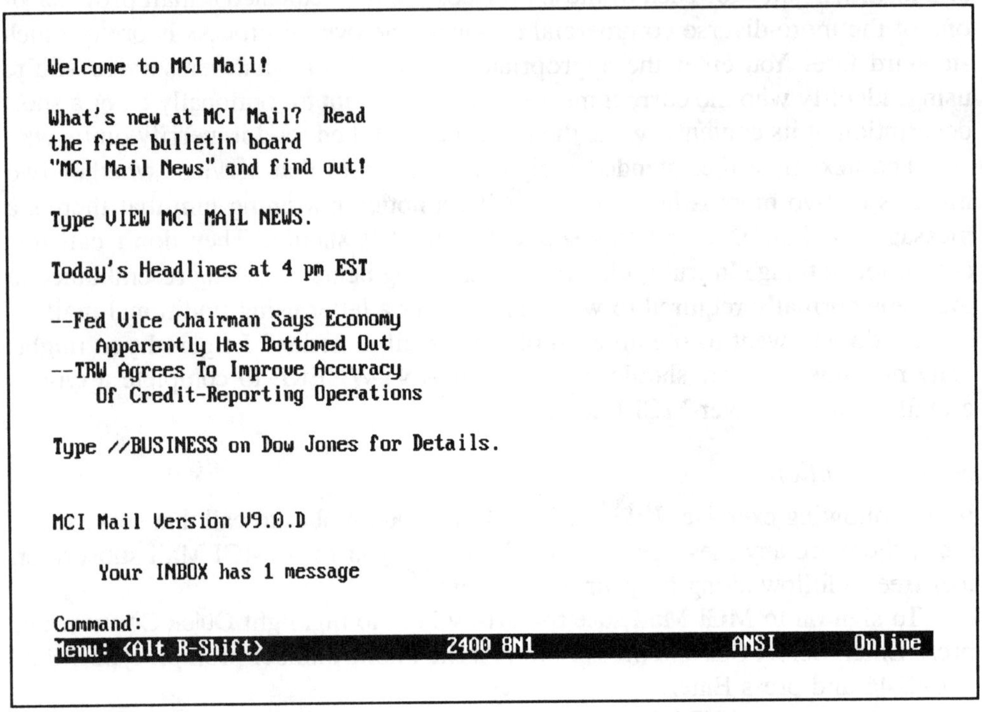

10-19 Most commercial services inform you when you have electronic mail waiting.

To scan your MCI INBOX, type SCAN INBOX and press Enter. This causes MCI to display the short message header shown in Fig. 10-20, informing me who wrote this message, when it was delivered, and a brief description of what it's about. Before reading this message, I want to show you another useful pcANYWHERE feature, one that can guarantee that electronic mail gets delivered to your PC.

To record this portion of your MCI Mail session, press Ctrl–Right Shift (or the appropriate hot-key command) to display the Terminal Options menu. Select Recording On and press Enter. At the next prompt, press Enter to specify that this session recording should be stored in AWR, the default path.

Your screen should now resemble Fig. 10-21, which shows pcANYWHERE asking you to identify the file in which you want to store the record you'll be keeping of the current session. For this exercise, I'll create a file called MCIMAIL.

To create a new session record file, press Ins. When pcANYWHERE asks

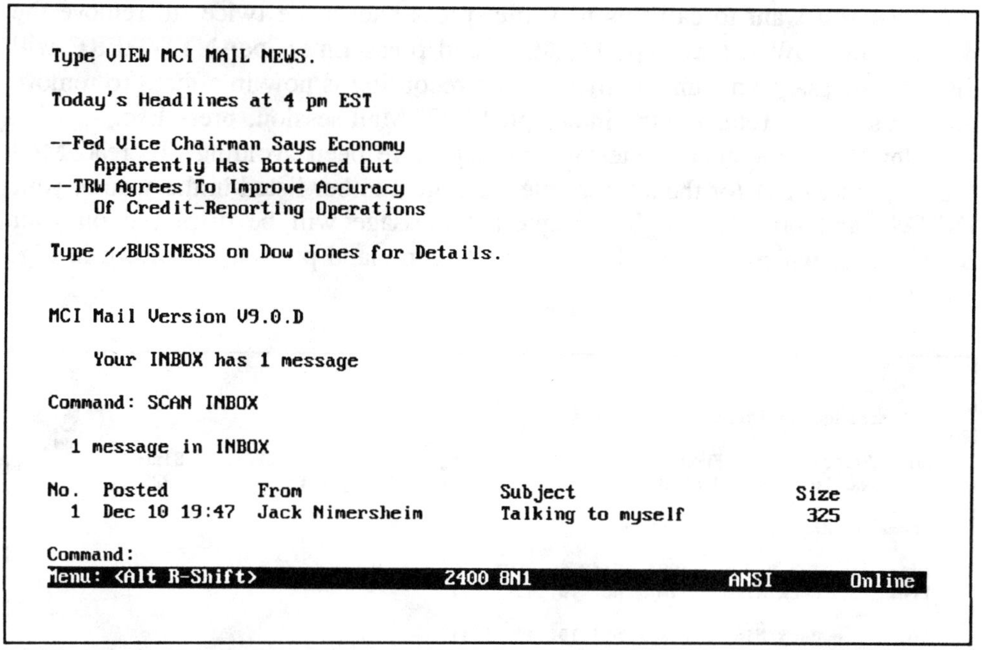

10-20 You can scan your MCI Mail INBOX to determine what messages are waiting for you.

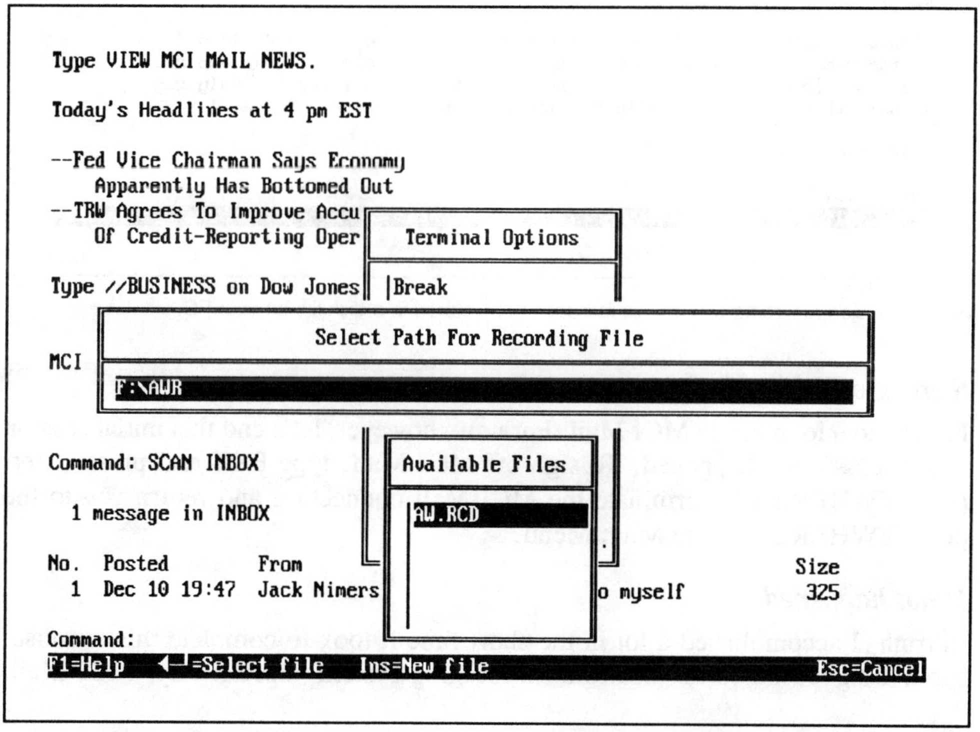

10-21 pcANYWHERE's Session Recording feature can be used to "deliver" mail to your PC.

you what you want to call this new file, press backspace twice (to remove the default name, AW), then type MCIMAIL and press Enter. pcANYWHERE will display a message announcing that session recording is now in effect. To remove this message and return to the interrupted MCI Mail session, press Esc.

Now let's read that mysterious message I've been avoiding. To read MCI Mail, type READ 1 (or the appropriate message number listed in the scan of your INBOX) and press Enter. The requested message will be displayed on your screen, after which MCI Mail returns to its command prompt, as shown in Fig. 10-22.

```
  1 message in INBOX

No. Posted        From                    Subject                    Size
  1 Dec 10 19:47  Jack Nimersheim         Talking to myself           325

Command: READ 1

Date:     Tue Dec 10, 1991  7:47 pm  EST
From:     Jack Nimersheim / MCI ID: 494-1341

TO:     * Jack Nimersheim / MCI ID: 494-1341
Subject:  Talking to myself

Jack,

They say you're in trouble when you start talking to yourself. I wonder
if the same rule applies to sending electronic messages to your own MCI
mail box. In my defense, I'll simply suggest that this was the only way
I could think of to logically demonstrate the advantages of E-mail.

Take care...um...Jack

Command:
Menu: <Alt R-Shift>            2400 8N1 REC            ANSI         Online
```

10-22 Session recording causes this message to be stored in a disk file on your PC.

Signing off MCI Mail

I could do a lot more in MCI Mail. For now, however, let's end this initial session and discuss what happened. To sign off MCI Mail, type EXIT and press Enter. pcANYWHERE will terminate the MCI Mail connection and return you to the pcANYWHERE/Remote Main Menu.

What happened

In truth, I accomplished a lot in the short time it took to complete this exercise. First, I signed on to an electronic messaging system—specifically, MCI Mail.

Next, I scanned my MCI Mail INBOX to determine what messages were waiting there for me. And finally, I read one message that was. Not only that, but I asked pcANYWHERE to deliver this message to my PC, thus making it available later if I want to review it. (And, yes, to answer what I know is your next question, I will, in chapter 13, *The pcANYWHERE utilities*.)

That's electronic messaging. It's quick, convenient, efficient, economical and, all things considered, relatively easy to use. Specific commands might differ from system to system, but the general procedures used to exchange e-mail on almost any commercial service or BBS resemble those in the previous exercise. You'll undoubtedly use pcANYWHERE for a lot of on-line activities, but I'd bet my last dollar that electronic messaging quickly becomes one of the most important.

Real-time conferencing

If using electronic mail is functional, perhaps the best way to describe real-time conferencing is that it's just plain fun. Sure, chatting with other people on-line can be practical, sensible, pedantic, and have a dozen other pragmatic purposes, but the true enjoyment associated with real-time conferencing is the simple act of meeting and getting to know new friends who drop by occasionally to visit the ever-growing on-line fraternity.

On-line line meetings can be informal—as would be the case, for instance, if you signed on to your favorite commercial service or BBS and noticed that someone you'd been meaning to talk with was there at the same time—or more structured. I and several writers, for example, gather in the Delphi Writers Group every Thursday night to discuss our shared interests.

Certainly, we discuss professional matters—which publishers are interested in specific types of projects, what editors have recently moved to a different magazine, how to put together an effective proposal, and the like. If you press me for total honesty, however, I'd have to admit that our primary purpose is to have a good time. One-liners pass through the phone lines like cash through the Federal budget. And we're just as likely to discuss puns as punctuation. (If you don't believe me, drop by some Thursday night at 9:30 pm; all are welcome.)

As is true with electronic messaging, almost every commercial service and even many local BBSs offer some form of conferencing. Once you begin participating in real-time conferences, you'll quickly get hooked. And as you become more familiar with this particular activity, you'll quickly come to realize that, like any community, the fraternity of on-line friends possesses its own unique language and social structure—not that you need to be overly concerned with any of this. Most conference groups will welcome you with open arms, but some idea of what to expect can help you feel more comfortable, should you decide to join in the fun.

Knowing the on-line language

If telecommunications has one down side, it's that it's a pay-as-you-play proposition. Whenever you call your favorite commercial service, it's a sure bet that someone, somewhere, is charging you something. Regular visitors to these electronic forums have come up with a number of creative ways for limiting the costs of their on-line activities, primarily by reducing either the time or the number of characters required to express common words, phrases, and the like.

Down through the years, a veritable dictionary of special abbreviations and acronyms has grown up around this effort. When you first start going on-line, these various terms and phrases will sound like a foreign language. And so, to help you break the ice, allow me to present "A course in on-line slang"—an informal lesson, if you will, covering some of the more esoteric axioms, idioms, and buzz words you're liable to encounter in a real-time conference.

ASAP ("As soon as possible") This familiar acronym, which pops up regularly in electronic messages or during on-line conferences, provides the perfect example of why such slang has invaded telecommunications. ASAP communicates with four letters an idea that otherwise would require a 20-character entry. When access charges are ticking away, such economic expressions can save a lot of money.

. . . (ellipsis) Used to indicate that the sender plans to continue his or her remarks in a subsequent entry. Most services limit individual entries to approximately two lines of text, after which you must press the Enter key and transmit your comment to the conference. If you're trying to communicate a thought that exceeds this length, ending one comment with an ellipsis tells people that additional text relating to the current entry will follow.

BRB (be right back) Used whenever someone temporarily departs an on-line conference. When other participants in the conference see this abbreviation, they know not to address comments to or expect responses from that individual until he returns.

BTW (by the way) Used just as it would be in normal conversation, as in, "BTW, I chatted with Bob yesterday and he's feeling better."

FYI (for your information) Yes, this inane abbreviation has crossed over from Corporate America to telecommunications.

GA (go ahead) A form of on-line etiquette. Commonly used at the end of a multiline entry—see the ellipsis code, above—to inform the other participants that the writer has finished his or her entire thought. Another common use for the GA code is when several people are attempting to type at the same time, which usually results in confusion. In this situation, one or more individuals will generally hold their own thoughts in abeyance and defer to a fellow participant by typing GA.

<grin> or <g> Used to end a comment offered in jest. Because people chatting on-line don't enjoy the luxury of being able to analyze visual or verbal signals, clues we humans rely on quite heavily to interpret face-to-face conversations, using this code helps ensure that others don't misinterpret your remarks. So as not to offend someone with a humorous remark, for example, you might end that remark with the <grin> code, as in, "Jane, is that garbage you're transmitting line noise, or are you studying a foreign language in night school? <grin>" BTW, you should always enclose a grin code in greater than/less than symbols (< >), as shown in the previous example, so that people realize it's an aside rather than part of your actual comment.

GMTA (great minds think alike) A quick way of responding to someone who's stated a position with which you agree. I've been on-line sometimes, madly typing in a response to something someone else said, only to find that another person has transmitted exactly the same thought just before I hit my Enter key. Situations like these are the ideal place to drop in a GMTA comment.

IMHO or IMO (in my humble opinion) Again, this comment is used just as it would be in normal conversation, with one caveat. As Michael Banks, a good friend and fellow writer, once pointed out to me, people who rely on this phrase overly much while participating in on-line conferences tend to deliver their messages while precariously perched atop some type of soapbox. Beware the IMHO people; it's not uncommon for them to confuse their humble opinions with arrogance.

LOL (lots of laughter, or laughing out loud) If someone says something funny or entertaining during an on-line conversation, you might want to let them know you appreciate their wit. A simple LOL should suffice.

OTF (on the floor) The highest on-line compliment you can offer someone in response to a humorous remark. Use this shorthand notation instead of LOL should someone enter something that really tickles your funny bone.

In addition to these basic slang terms, on-line veterans also rely on special characters sequences to clarify or, in some cases, simply embellish the appearance of their electronic repartee. One of the most common applications of this technique is if you want to direct a comment at a specific member of a real-time conference. To accomplish this, precede your comment with a greater than symbol (>) pointing to that person's name, as illustrated in, ">RALPH, I couldn't agree more."

A simple variation on this theme is to add several dashes before the greater than symbol. This technique is especially useful when the service itself uses greater than symbols to identify individual speakers, as explained in the next paragraph. Adding a series of dashes under these circumstances guarantees that person to whom you are directing a comment recognizes your intentions, as in, "NIMS>---> RALPH, I couldn't agree more."

As I mentioned in the previous paragraph, some services distinguish among the different participants in a real-time conference by automatically preceding each on-line comment with the name (or *handle*) of the person who entered it. Flip-flopping the previous notation, therefore, provides a simple way for identifying a remark you want to make about yourself. One example of how this technique might be used is, "NIMS>---feeling beat tonight."

A few paragraphs earlier, I pointed out how conference participants often use a <grin> message to respond to a comment designed to elicit a laugh from other users. A more picturesque variation on this same theme is to end your comic comment with a colon, followed by a right parenthesis, as in, "I notice Bill didn't sign on tonight. His phone bill must have arrived today. :)" What does this cryptic symbol mean? Cock your head to the left and take another look at it sideways. Now do you understand? You can even add a wink, as in ;).

There you have it, a course in on-line slang. While no one will denigrate you should you elect not to use these various terms and techniques when you sign on to a real-time conference, a basic familiarity with the unique language of on-line communications can help avoid confusion should one or more of them surface during a real-time conference in which you're participating.

On-line etiquette

For all of its positive attributes, telecommunicating has one major shortcoming; it tends to be extremely one-dimensional. When carrying on a traditional conversation, we often rely on more than words to communicate. For example, whether by design or instinct, almost all of us are able to interpret body language, which in turn helps us analyze the true meaning behind the words being said. Vocal inflection also provides valuable clues to someone's intent when they engage in a face-to-face discussion.

None of these subtle clues are available in an electronic conversation. Consequently, you have to be extra careful that the purpose of your prose is fully understood. That's one valuable function of the various shorthand expressions introduced in the previous section.

A second potential peril associated with the impersonal quality of real-time conferencing is a tendency to forget that the people on the other end of a connection are also real, live individuals with feelings and, in some cases, problems or fragile egos of their own. I've watched people I know well take on an entirely new and not very endearing personality in the impersonal environment of a real-time conference.

Please don't misunderstand. I'm not trying to inhibit you with these observations. The best way to enjoy yourself in a real-time conference is to express yourself freely and completely. A little empathy and diplomacy can go a long way, however, toward making that conference an enjoyable experience for everyone involved. One technique I've found helpful is to look upon being welcomed into an on-line conversation as the electronic analogy of being invited to a party. If you're considerate of and courteous to your fellow guests, they'll extend the same consideration and courtesy to you.

But the best way to learn about real-time conferences is to participate in one. As I said earlier, feel free to stop by our Writer's Group conference some Thursday night. How do you get there? That's easy. Simply follow the following steps after you sign on to Delphi and are presented with that service's opening menu (see appendix A, *What's out there?*, for information on how to contact Delphi and inquire about subscribing to this popular on-line service):

1. Type GROUP WRITERS and press Enter.
2. Type CONFERENCE and press Enter.
3. Type WHO and press Enter.

The previous step will display a listing of everyone participating in our weekly get-together on a particular Thursday night, along with a conference number Delphi has assigned to that evening's festivities. To join us, type the conference number displayed in response to your WHO command and press Enter.

We'll be notified that you've entered our little group and you'll probably see a number of personal welcome messages from the various people signed on that night. Then simply sit back and join in. Type in any comments, questions, observations, etc. you have that relate to the topic of the night—and bring along your best puns, just in case there's no "topic of the night" that night. Pressing Enter will cause your words to magically appear on our screens. If you have questions about any Delphi commands or procedures, feel free to ask them also. Someone will graciously provide assistance, usually our Delphi expert-in-residence and fellow TAB author, Dick Evans. First and foremost, however, relax and have fun. That, after all, is what real-time conferencing is all about.

Summary

Like I said at the beginning of this chapter, I could spend hours discussing the joys of telecommunicating. I hope these few examples of what's available on the thousands of commercial services and local BBSs across the country—indeed, around the world—have piqued your interest enough to experiment with more than just the admittedly impressive remote-computing features of pcANYWHERE. Your efforts will be amply rewarded, I can assure you.

11 CHAPTER

Automating your on-line activities

Automation and computers have always been associated with one another—two sides of the same electronic coin. Say *computer*, and people think *automation*. Say *automation*, and they conjure up an image of some type of computer. The relationship is that unequivocal.

pcANYWHERE provides two methods for automating remote communications—automated procedures and scripts. In this chapter you're going to examine how these two pcANYWHERE features can be used to transform going on-line into a hands-*off* experience, to put a new spin on an old expression.

Procedures

pcANYWHERE's Automated Procedures option provides a "quick and dirty" way to automate simple operations between a host and remote system. In theory, creating a procedure is similar to programming a video recorder to tape a specific television show at a predetermined time. In practice, thankfully, Symantec has gone to great lengths to make setting up a new pcANYWHERE procedure much less difficult than using a VCR.

Personal note: I, for one, am extremely grateful to Symantec for its efforts in this area. Many's the time I've clicked on my VCR, expecting to see one program, only to discover that I'd inadvertently included the wrong time or channel in my taping instructions. Nothing is more frustrating than sitting down to watch that much-awaited episode of *Star Trek: The Next Generation* which a pressing deadline forced me to miss and finding, instead, a one-hour excerpt from a heated Senatorial debate on the socio-political ramifications of purchasing scented soap for the Congressional restrooms, or some such equally insipid drivel. It's always baffled me how someone capable of programming a computer to do everything but take out the garbage can be so inept at running a piece of electronic equipment even my five-year-old son has managed to master.

What types of tasks can you automate with the pcANYWHERE Automated Procedures features? In addition to file transfers, which procedures handles easily, you can also execute single-line DOS commands up to 40 characters long on either the host or remote system. In other words, anything you can accomplish with a DOS command less than 40 characters in length can be automated with a pcANYWHERE procedure. This includes running any of the standard DOS commands (DIR, DEL, COPY, BACKUP, and the like) or starting specific application programs. If pcANYWHERE's 40-character limitation proves too restrictive, you can often work around it with a well-designed batch file. Practical uses for procedures include:

- Initiating unattended file transfers between a host and remote system late at night, thus taking advantage of reduced off-hour long-distance rates.
- Automating complex DOS operations by using a procedure to execute a batch file on either the host or remote system at a predetermined time.
- Using macros or other programming capabilities built into a specific application to perform complicated tasks remotely, automatically, and unattended, at predetermined times.
- Scheduling daily maintenance on a host or remote system, which can incorporate such activities as archiving important data, deleting outdated files, defragmenting a hard disk, and the like.

The following exercises are designed to demonstrate how the pcANYWHERE Automated Procedures feature might be used in a familiar situation.

Putting procedures into perspective

Let's take a moment and try to put pcANYWHERE into perspective—specifically, the perspective of being a valuable business tool that a company relies on to coordinate the activities of its employees. This hypothetical company, which I'll call Widgets, Inc., sells (logically enough) widgets. In an effort to decrease the turnaround time required to fill and ship customer orders, the President of Widgets, Inc., decided that she needed to find some way for her six regional sales managers to take each day's orders and electronically transmit them to a PC located in the company's central headquarters.

Enter pcANYWHERE. Even more correctly, enter the pcANYWHERE Automated Procedures feature.

The President began by installing pcANYWHERE/Host on the PC to which these daily sales reports would be sent. She then purchased additional copies of the pcANYWHERE/Remote package, one for each of the six regional sales managers. Finally, the President called her managers and walked them through the following exercise, with one variation for each manager, thus setting up their systems to automatically transmit the desired reports overnight, every night, so that they would be waiting on the host system when she arrived at work the next morning.

Note: The variation for each sales manager is a single item, the time of the call. This ensures that two remote systems won't try to call the host at the same time. Except for these two minor differences, the steps required to create an automated procedure for each of the remote systems is identical.

I'll be using the MY HOST listing on your remote computer, which you created in chapter 7, for the following exercise. MY HOST is already set up to automate the process of signing on to your host. This is a critical requirement for using an automated procedure that involves unattended host and remote systems, given that no one will be available at either computer to manually enter the proper sign-on sequence.

Building an automated procedure

To automate file-transfer with an automated procedure, use the arrow keys to highlight Automated Procedures on the pcANYWHERE/Remote Main Menu and press Enter. This will display the option box shown in Fig. 11-1, which is used to list the names of any automated procedures set up on your remote computer. Because this is the first one you're creating, no entries show up in Fig. 11-1. To create a new procedure, press Ins. When pcANYWHERE asks you to name your new procedure, type Transmit daily sales report and press Enter.

At this point, you'll be presented with the option of either executing or viewing/modifying the selected procedure, as shown in Fig. 11-2. It's somewhat difficult to execute a procedure that doesn't exist yet, so select the View/Modify option. This will allow you to define the individual steps required to have your remote call the Widget, Inc., host and upload the appropriate file.

Select View/Modify the Procedure and press Enter. This will display a window containing a list of hosts associated with the selected procedure. (It's possible to use the same procedure to automate activities for multiple hosts, something I'll go into more in a few pages.) Once again, because you're creating a new procedure, this list will be blank.

To add a host to your sales report procedure, press Ins. pcANYWHERE will display a window containing the various host listings on your remote system. To

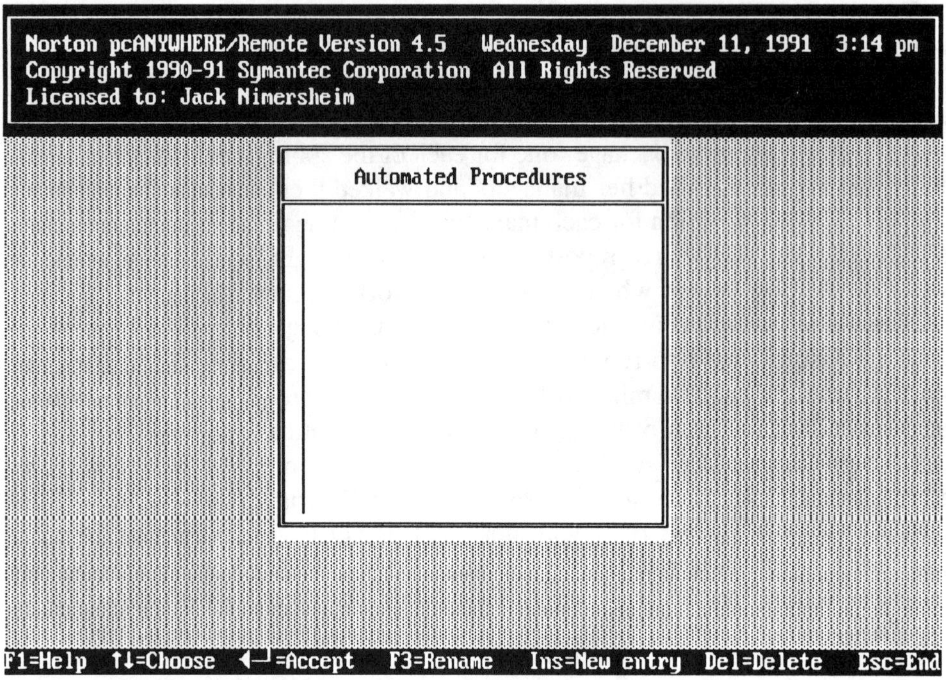

11-1 Automated procedures let you perform repetitive tasks with a single command.

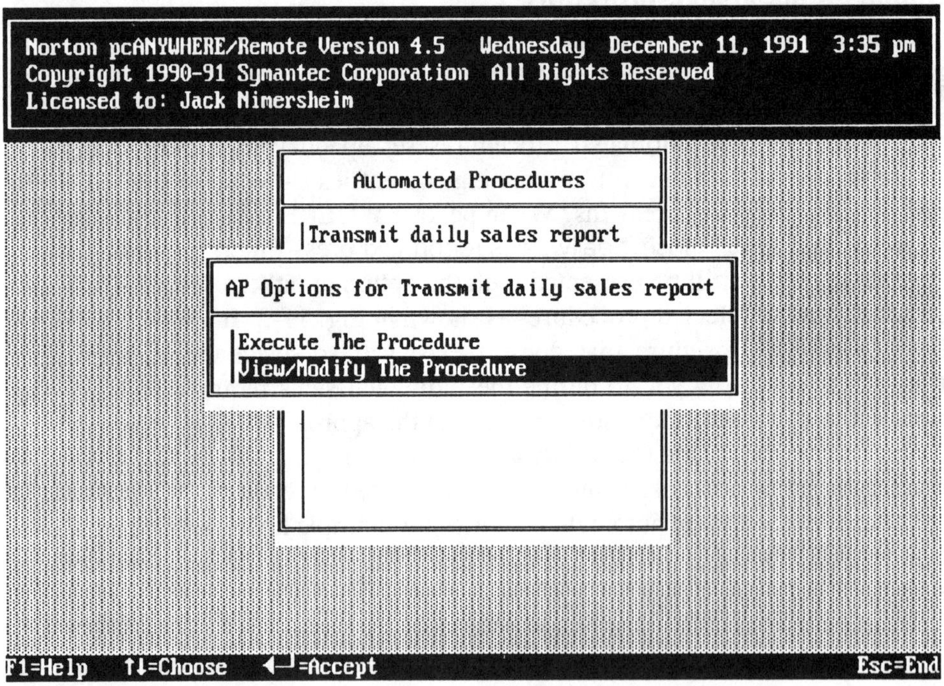

11-2 You use the View/Modify option to create a new procedure.

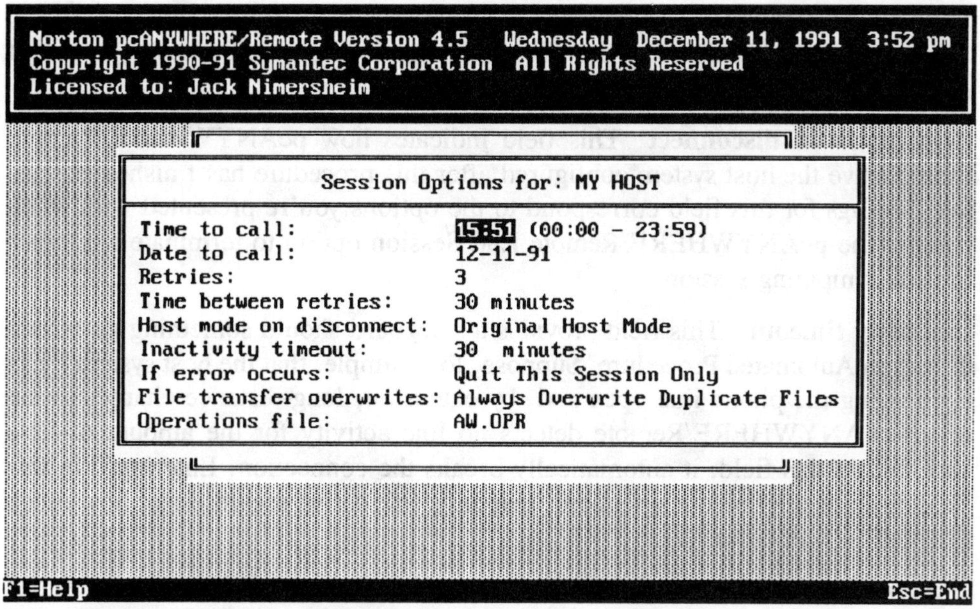

11-3 The Session Options dialog box lets you describe when and how a procedure should be executed.

select the host listing you created in chapter 7, use the arrow keys to highlight MY HOST and press Enter.

After you select a host, pcANYWHERE will advance you to the Session Options dialog box shown in Fig. 11-3. Use this dialog box to enter information pertaining to the session during which you'll run this procedure for MY HOST, the host system selected in the previous step. The session information pcANYWHERE needs to coordinate your automated activities includes:

Time to call Use this field to indicate what time you want pcANYWHERE to run the current procedure for this host. Because a procedure can be executed at any time, day or night, you must enter the time in 24-hour, military format, where afternoon and evening hours—that is, the time between noon and midnight—are indicated with the numbers 13-24 (e.g., 1:00 P.M. equals 13:00, 2:00 P.M. equals 14:00 hours, etc.). Entering nothing in the time field will cause a procedure to be executed immediately.

Date to call Use this field to indicate the date on which you want this procedure executed, using the format MM-DD-YY. Leaving the date field blank will cause pcANYWHERE to execute the procedure every day.

Retries Use this field to specify how many times pcANYWHERE should attempt to run this procedure if the host system is busy when the first call is placed. You can specify any number of times less than 100 (0 through 99).

Time between retries Use this field to specify the number of minutes, up to 99, that pcANYWHERE/Remote should wait before trying to call a host system that was busy.

Host mode on disconnect This field indicates how pcANYWHERE/Remote should leave the host system configured after this procedure has finished executing. Settings for this field correspond to the options you're presented with when you use the pcANYWHERE/Remote End Session option to terminate any other remote-computing session.

Inactivity timeout This field provides a safeguard should something go wrong during an Automated Procedure. Suppose, for example, that the host system locks up, leaving the phone line open and, by extension, long-distance charges piling up. If pcANYWHERE/Remote detects no line activity for the amount of time specified in this field, it automatically breaks the connection. Entering 0 in this field disables the timeout safeguard, leaving a line active regardless of how much time passes without activity.

If error occurs Use this field to indicate what action pcANYWHERE/Remote should take if an error occurs while the procedure is being executed. Potential errors that can trigger this action include an inactivity timeout, a failed file transfer, a malfunctioning application program, and the like.

File transfer overwrites Use this field to indicate the action that pcANYWHERE should take if, during file transfer, it encounters a file with the same name on the target system.

Operations file Use this field to identify the file containing the specific actions pcANYWHERE should perform whenever you run this automated procedure.

With this information as background, let's go ahead and set up the automated procedure for Widgets, Inc.:

1. Highlight the Time to Call field and press Enter.
2. Type 02:00 and press Enter. (This tells pcANYWHERE to run this procedure at 2:00 a.m.)
3. Press Del until the Date to Call field is empty, and press Enter. (This inserts TODAY in the date field.)
4. Use the arrow keys to highlight File Transfer Overwrites and press Enter.

pcANYWHERE will present you with three options:

- Always Overwrite Duplicate Files
- Never Overwrite Duplicate Files
- Overwrite Older Files Only

Clearly, in your imaginary scenario, you'd want to guarantee that only the newest sales report makes it onto Widget's host system. So use the arrow keys to highlight Overwrite Older Files Only and press Enter. pcANYWHERE will advance you to the Operations File field. This is where the real work is done when you first create an automated procedure.

To indicate the specific actions pcANYWHERE should perform whenever you run the sales report procedure, press Enter to select an operations file. When the Select Path prompt appears, press Enter to accept C:\AWR, your pcANYWHERE directory. pcANYWHERE will display an option box listing all available operations files in the specified directory. Currently this box contains one file, AW.OPR, a default operations file. You could use this file, but why don't you create a new operations file with a name that provides some clues as to what it accomplishes?

To create a new operations file, press Ins. When the Enter File Name prompt appears, press backspace to erase AW (the default name), type REPORT, and press Enter. pcANYWHERE will advance you to its Operations dialog box, shown in Fig. 11-4—just another empty window at this point. Don't let appearances fool you, however. With a few simple keystrokes you can fill this window with a list of instructions you want pcANYWHERE to execute whenever the new REPORT procedure is run. How simple? Pay attention and you'll see.

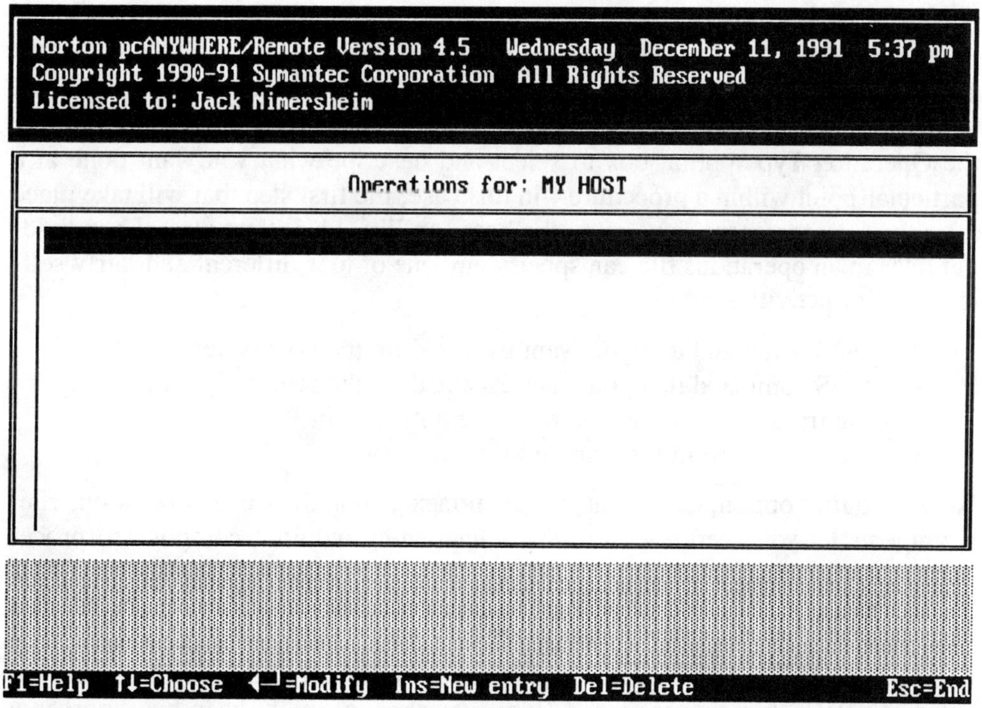

11-4 Procedures consist of a series of command lines.

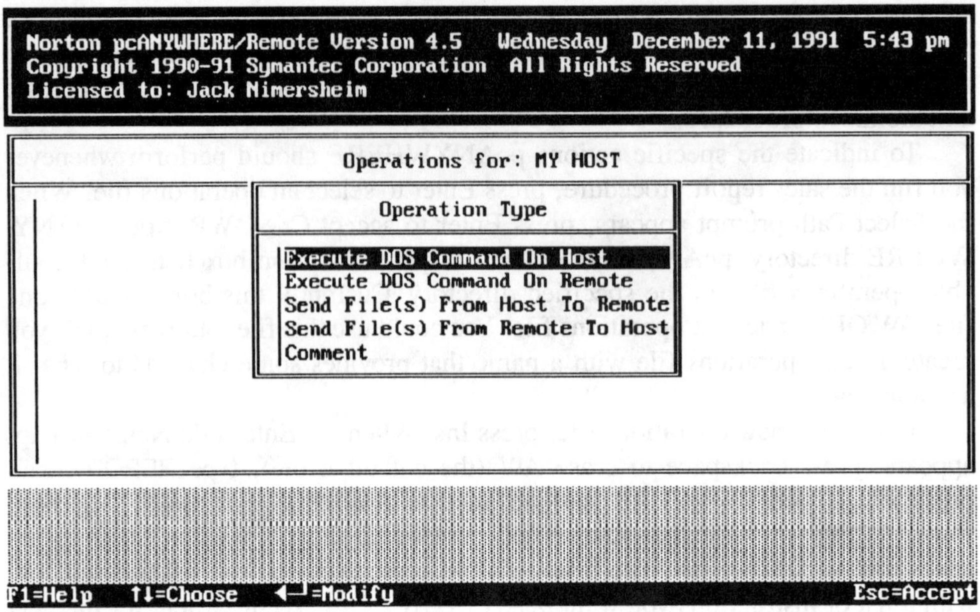

11-5 You use the Operation Type menu to indicate what type of activity you want to perform in a procedure.

To select the first operation you want performed when you run the REPORT procedure, press Ins. This displays the Operation Type option box shown in Fig. 11-5. An operations file consists of one or more lines, each of which describes a specific activity that you want performed on either the host or remote system. Use the Operation Type option box to select and describe what you want done at a particular point within a procedure—in this case, the first step that will take place after the remote system signs on to the host. As Fig. 11-5 illustrates, the individual lines in an operations file can specify any one of four different and fairly self-explanatory activities:

- a DOS command that you want executed on the host system
- a DOS command that you want executed on the remote system
- a file transfer from the host to the remote system
- a file transfer from the remote to the host system

Note: The fifth option, Comment, inserts nonexecuting lines in an operations file. Comments let you keep track of what's happening within a complicated procedure.

For this exercise, let's assume that each sales manager's daily report will be a file stored in a directory called REPORTS on drive C of the remote system. The name of this file will be unique for each manager, but all of them will use the file extension RPT. A manager named Bob, for example, might keep his report in a file named BOB.RPT. Another manager, Joan, could call hers JOAN.RPT. (The secret here is to make sure no two managers use the same filename. Unique names eliminate the possibility of one report overwriting a previously submitted report with the same filename.) Your final assumption is that the President of Widgets, Inc. wants to transfer these daily reports to a directory on the host system called DAILIES. With these assumptions in mind, let's go ahead and build an operations file that will transfer each sales manager's report from his computer to the host system.

To build this operations file, highlight Execute DOS Command on the Remote and press Enter. pcANYWHERE will display the operation prompt, shown in Fig. 11-6. This is where you enter the DOS command you want executed first each time this procedure is run. Use this first command to make certain that drive C on the remote system is the active drive.

To make drive C active, type C: and press Enter. pcANYWHERE then asks you to verify that you want to add this command to your operations file. You do. To add this command to the operations file, highlight Yes and press Enter.

This returns you to the Operations dialog box, as shown in Fig. 11-7. Notice that the command you entered in the previous step has been added to your operations file, in the format:

REMOTE DOS: C:

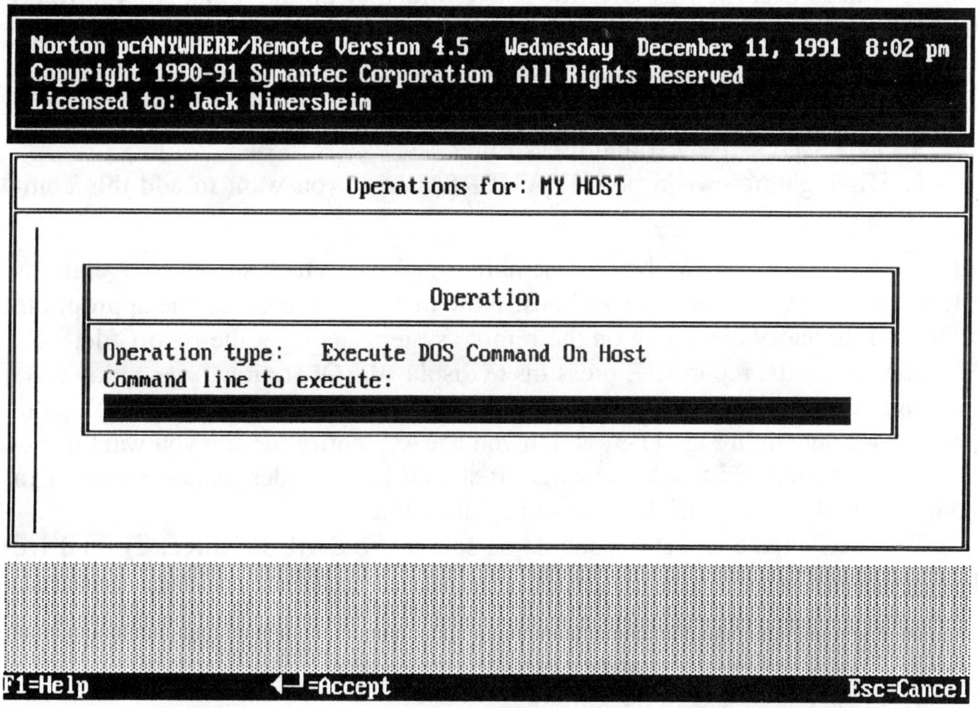

11-6 You can specify the DOS command you want executed on either the host or remote computer.

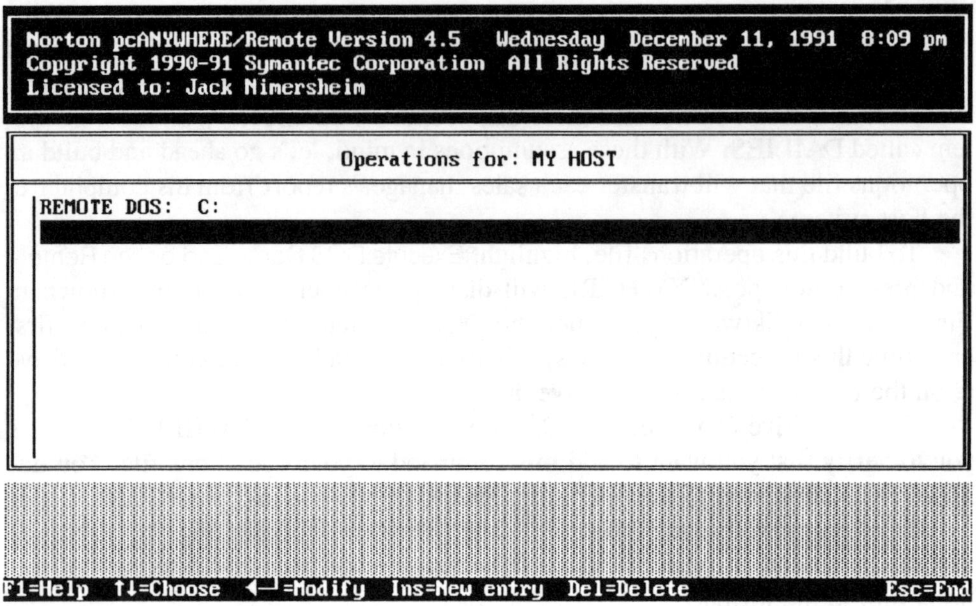

11-7 pcANYWHERE automatically adds the appropriate command line to your procedure.

Now you need to make REPORT the active directory on the remote drive. To do this:

1. Press Ins to display the Operation Type option box.
2. Highlight Execute DOS Command on the Remote and press Enter.
3. Type CD \REPORT and press Enter.
4. Highlight Yes when pcANYWHERE asks if you want to add this command, and press Enter.

At this point, your display should resemble Fig. 11-8, which shows an operations file that currently consists of two lines. Now that you've ensured the appropriate drive and directory are active on the remote system, let's get the report file.

To transfer the report file, press Ins to display the Operation Type option box. Highlight Send File(s) from Remote to Host and press Enter. This will display the prompt box shown in Fig. 11-9, which you use to identify the file you want transferred (source name), as well as where it should be sent (destination name). I'm going to employ a little trick here, so pay attention.

To specify that a report file should be sent to the DAILIES directory on drive C of the host system:

1. Type *.RPT and press Enter.
2. Type C:\DAILIES and press Enter.
3. Press Esc to accept these values.

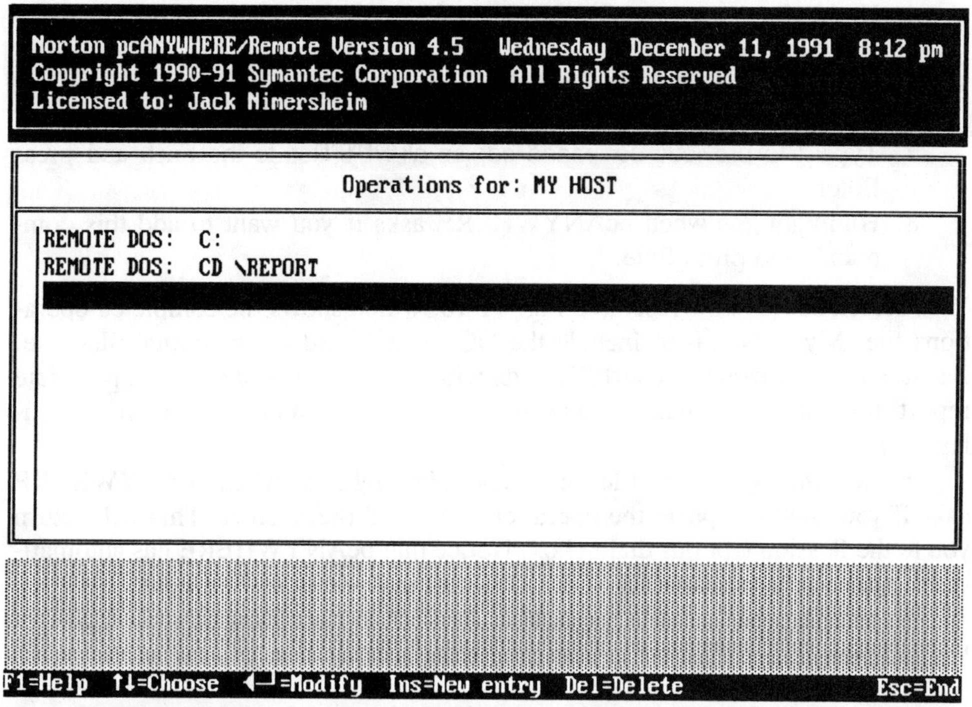

11-8 Each different activity you specify becomes a new line in your procedure file.

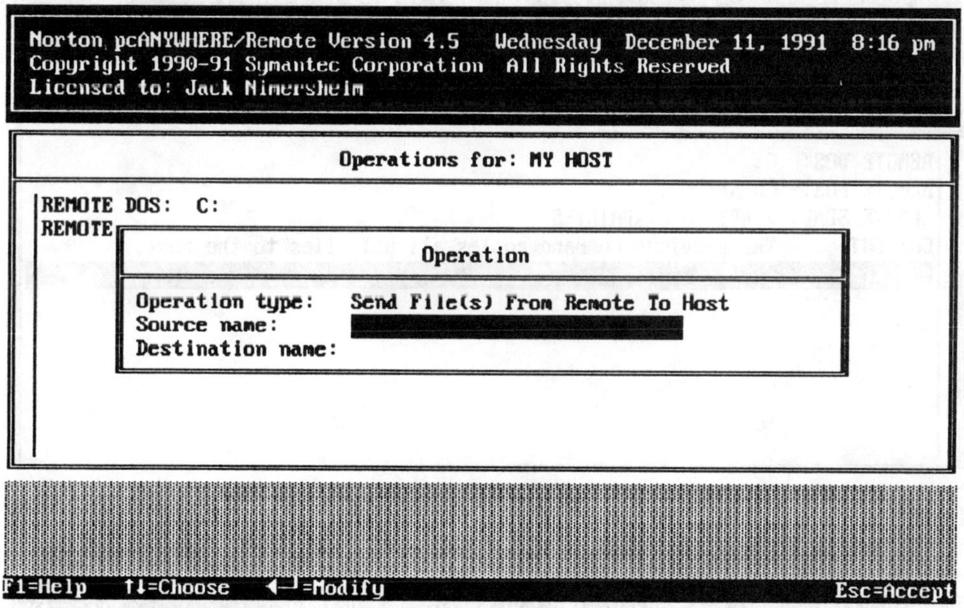

11-9 Procedures can also be used to exchange files between the host and remote systems.

4. Highlight Yes when pcANYWHERE asks if you want to add this command and press Enter.
5. Press Ins to display the Operation Type option box.
6. Highlight Comment and press Enter.
7. Type The previous command copies all RPT files to the host. and press Enter.
8. Highlight Yes when pcANYWHERE asks if you want to add this command, and press Enter.

Your screen should now resemble Fig. 11-10, which shows the completed operations file. My trick was to include the DOS * wildcard in my source filename. Because each manager's report file ends with RPT, this transfers the appropriate report, regardless of which remote system is connected to the host at any particular time.

To save this operations file, press Esc. Highlight Yes when pcANYWHERE asks if you want to update the operations file, and press Enter. This will return you to the Session Options dialog box. Notice that pcANYWHERE has automatically inserted the filename REPORT.OPR in the Operations File field.

To save the new host file, press Esc. At the Add New Host prompt, highlight Yes and press Enter. This will return you to the host window you saw earlier. This

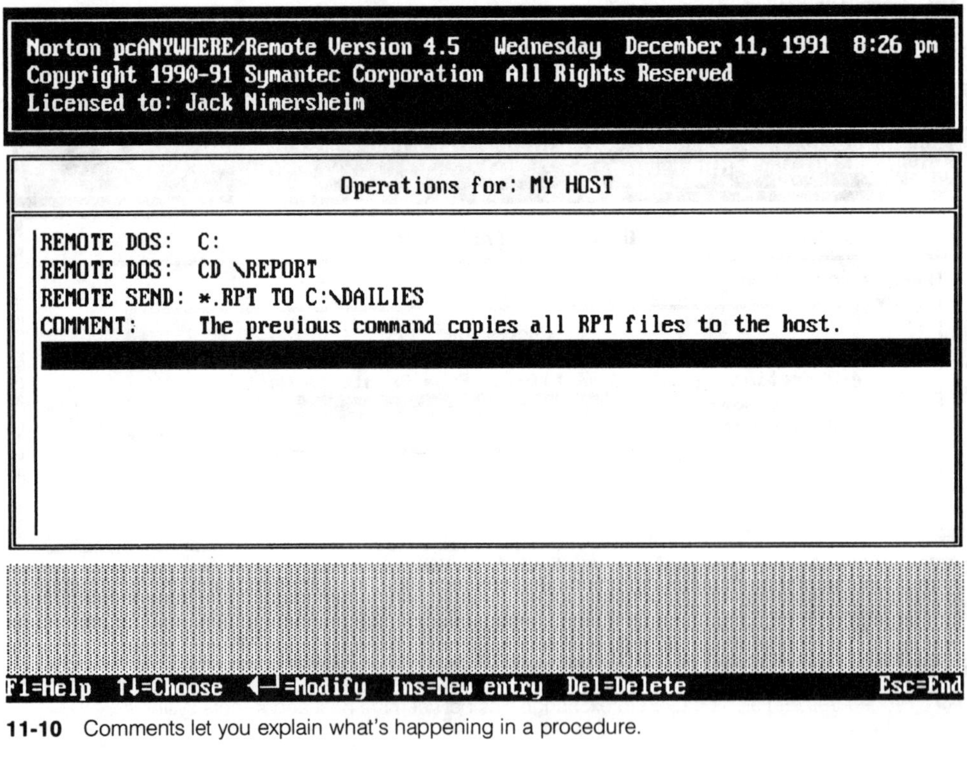

11-10 Comments let you explain what's happening in a procedure.

time, however, instead of being empty, it includes a listing for MY HOST, indicating that this procedure, when selected, will be executed for that host system.

To exit this host window and return to the Automated Procedures options box, shown in Fig. 11-2, press Esc. Because you're not set up to run this procedure—neither the directory nor the file it's designed to access and transfer exist on your remote system—there's no sense in trying to execute it. For now, therefore, press Esc twice to return to the pcANYWHERE/Remote Main Menu.

Executing a procedure

By repeating these steps on each sales managers' remote system, the President of Widgets, Inc., could easily put into place the policy she desires—that is, transferring each manager's sales report to a central PC every night. And how would the sales managers make sure this happened? I'm glad you asked. All they'd need to do is execute the new procedure, using the following steps:

To execute an automated procedure, use the arrow keys to highlight Automated Procedures on the pcANYWHERE/Remote Main Menu and press Enter. When the Automated Procedures listing appears, highlight Transmit Daily Sales Report and press Enter. At the AP Options box, highlight Execute the Procedure and press Enter.

And what does the manager need to do now? In a word, nothing. He can go to bed and pcANYWHERE will handle everything else, automatically calling the host at the prearranged time and transferring his report file under the control of the REPORT.OPR file.

Tip: If the imaginary President of Widgets, Inc., really wanted to be tricky, she could create the REPORT.OPR file herself, using her own copy of pcANYWHERE, then transfer a copy of it to each of her regional manager's computers. After all, only the time of the transfer, recorded in each Session Options dialog box, varies for individual managers.

This simple scenario should give you some idea of the potential uses for pcANYWHERE's Automated Procedures feature. And keep in mind that it's the lesser of pcANYWHERE's two methods for automating your on-line activities. Wait until you see what a well-designed script can accomplish, which is exactly what you'll do in the next section.

The pcANYWHERE script language

Having just created a pcANYWHERE automated procedure, you already have a pretty good idea of what a script is. Basically, a script is a series of commands and procedures that execute automatically each time that script is run. A script can be as short and simple as a single command, or as long and complex as a given situation demands. Length and complexity, however, are the areas in which scripts excel.

A properly designed script can handle even the most complex on-line situations—totally on its own, and without user intervention. A pcANYWHERE script could be designed, for instance, to initially try connecting with CompuServe via a local access number. If this first attempt failed, however (as might be the case if all the local lines were already in use), the script could include additional instructions to go ahead and contact CompuServe using a secondary access channel, such as the Tymnet or SprintNet packet-switching network.

Scripts, then, let you automate more complex activities. It's possible, for example, to write a script that would call CompuServe, check to see if you have electronic mail waiting, switch to the CompuServe Mail area if any messages do exist (or, alternately, immediately disconnect from CompuServe if your electronic mailbox is empty), transfer these messages to an archive file on your hard disk, and end the current on-line session. Then, at your convenience, you could review this archive file in much the same way you currently open and read the letters, magazines and, of course, junk mail (which is not yet a major problem on-line) delivered to your home by the local letter carrier. You could even design this script to begin execution at a specified time each day—for example, late at night, when connect charges are less expensive—to totally automate retrieving electronic mail delivered to you over CompuServe Mail. That, briefly, is what a script is, what a script does, and what a script is capable of doing.

The next natural question, then, becomes, "What mechanism exists to support such impressive automation of on-line activities in pcANYWHERE?" The answer is the script language Symantec added to recently released version 4.5 of The Norton pcANYWHERE.

What is the Norton pcANYWHERE script language?

I keep offering these obvious definitions, but Norton's script language is a series of special commands that allow you to automate your pcANYWHERE operations. That about covers it. With the appropriate knowledge and any ASCII (standard text) editor, you can create script files that can then run at anytime within a pcANYWHERE session in order to automate a wide range of activities. pcANYWHERE supports such advanced script features as:

- Allowing user input during script execution
- "Chaining" scripts, where one script automatically begins execution of a second script
- Program branching to labeled subroutines
- Conditional branching, using standard IF/THEN/ELSE statements
- Redirection capabilities to, for example, redirect output to a system printer
- Execution of DOS commands from within a script file
- String manipulation and mathematical capabilities

Some pcANYWHERE script ground rules

Before you begin analyzing the individual commands that comprise the pcANYWHERE script language, I'll outline some simple rules governing the design of a pcANYWHERE script. There aren't many, and the ones that do exist are relatively straightforward. Specifically:

- pcANYWHERE scripts are created as standard ASCII files and then compiled for execution using the Preprocess a Script utility.
- Each pcANYWHERE script command must begin on a separate line.
- Each pcANYWHERE script command must be complete on a single line.
- Each command line consists of one or more reserved words (instructions) and, where appropriate, one or more *arguments* that clarify how the instruction is to be executed or, alternately, where the results of a command are to be stored.
- Arguments can consist of variables, constants, or arithmetic expressions.
- Commands and their parameters are not case-sensitive.
- You can include comments in a command line by placing them after a semicolon.

That's about it. Beyond following these seven simple rules, how you design your pcANYWHERE script and what you use it to accomplish is pretty much up to you. With this said, let's take a look at a sample pcANYWHERE script and analyze what it's designed to accomplish.

A sample pcANYWHERE script

Figure 11-11 shows a sample pcANYWHERE script. As this figure demonstrates, you can display a pcANYWHERE script with the DOS TYPE command. This is possible because you create pcANYWHERE scripts in ASCII format, a fact that also makes them relatively easy to design and revise with most popular word processors.

Using comments

Notice that the boxed message at the beginning of this particular script simply explains what it's designed to do. Each of the first four lines begins with a semicolon, so their contents are recognized by pcANYWHERE as being comments, rather than containing script commands it should execute.

This introductory information is followed by a blank line, which pcANYWHERE totally ignores. Blank lines provide a convenient way of making your script more readable by visually separating its different sections.

```
(F:\AWR) type cis.scr
;**************************************************************************
;* A sample pcANYWHERE script file to log onto CompuServe via local      *
;* CIS access number                                                      *
;*                                                                        *
;**************************************************************************

LET u = "xxxxx,xxxx"                   ; Insert your CompuServe User ID here
LET p = "yyy,yyyyy"                    ; Insert your CompuServe password here

EMULATE ANSI                           ; Select terminal emulation
LOAD CONFIG "Non-pcANYWHERE 2400"      ; Load a hardware configuration
DIAL NUMBER "555-5555"                 ; Insert your local CIS # here
WAIT CARRIER                           ; Looks for data carrier detect (CDC)
SEND CHAR 13                           ; Sends carriage return to activate CIS
WAIT STRING "Name:"                    ; Looks for system prompt
SEND LINE "CIS"                        ; Identifies CIS as desired service
WAIT STRING "ID:"                      ; Looks for CIS User ID prompt
SEND LINE u                            ; Sends User ID followed by CR
WAIT STRING "password:"                ; Looks for CIS password prompt
SEND LINE p                            ; Sends password followed by CR

(F:\AWR)
```

11-11 A sample pcANYWHERE script file.

Assigning variables in a pcANYWHERE script

Now consider the next two lines, which are actually the first command statements in the CIS.SCR script:

```
LET u = "xxxxx,xxxx"     ; Insert your CompuServe User ID here
LET p = "yyy,yyyyy"      ; Insert your CompuServe password here
```

These lines use the pcANYWHERE LET command to assign different text strings to a pair of variables. Specifically, your CompuServe user ID is assigned to variable u and your CompuServe password is assigned to variable p. Notice, also, that I've included comments on both of these command lines explaining their purpose.

As a rule, it's a good idea to define certain values as variables in this way near the beginning of a script. Then, within the actual command lines that script executes, you would include the appropriate variable name rather than the value itself. As a result, should a value associated with a given variable need to change later—as would be the case, for example, if, for security reasons, you decide to select a new CompuServe password—simply assigning a new value to that variable would update your entire script. (If this explanation confuses you, don't worry; you'll see what I mean more clearly as you delve a little further into the CIS.SCR script.)

Setting terminal emulation

Following a blank line, once again inserted to make the script more readable, is another pcANYWHERE script command:

```
EMULATE ANSI      ; Select terminal emulation
```

The EMULATE command is used here to identify which terminal type pcANYWHERE should emulate during any CompuServe session initiated with this script. In this case, an ANSI terminal is specified, a fairly common terminal emulation for on-line information service.

Loading host configurations in a script

The next line in your CIS script loads the appropriate hardware configuration into pcANYWHERE for contacting CompuServe. In this case, I've selected the non-pcANYWHERE configuration, which is set up for 2400-bps communications using no parity, the appropriate settings for CompuServe. As Fig. 11-11 illustrates, the LOAD CONFIG command selects the appropriate hardware configuration from within a script file, using the following structure:

```
LOAD CONFIG "Non-pcANYWHERE 2400"      ; Load a hardware configura
tion
```

Dialing a remote computer

Once your parameters are set properly, the next logical step is to actually dial the number of the remote computer you want to contact. The pcANYWHERE script command for accomplishing this is DIAL NUMBER. (That makes sense, doesn't it? See, I told you this stuff was relatively easy.) The following line in your sample CIS.SCR script accomplishes this:

```
DIAL NUMBER "555-5555"      ; Insert your local CIS # here
```

Of course, I don't know where you live, so I can't include the correct CompuServe access number for your location within this example. Recognize, therefore, that, should you decide to incorporate my sample script into your pcANYWHERE environment, you'll need to substitute the appropriate CIS local access number in the above command line before CIS.SCR will execute properly.

Receiving and transmitting data

Of course, once you reach the local CIS number, you'll still need to request access to CompuServe and enter your sign-on information before the actual on-line session can begin. To accomplish this, the CIS.SCR script contains the following command sequence:

```
WAIT CARRIER                 ; Looks for data carrier detect (DCD)
SEND CHAR 13                 ; Sends carriage return to activate CIS
WAIT STRING "Name:"          ; Looks for system prompt
SEND LINE "CIS"              ; Identifies CIS as desired service
WAIT STRING "ID:"            ; Looks for CIS User ID prompt
SEND LINE u                  ; Sends User ID followed by CR
WAIT STRING "password:"      ; Looks for CIS password prompt
SEND LINE p                  ; Sends password followed by CR
```

Let's examine this final command sequence: The WAIT CARRIER command tells pcANYWHERE to wait until a connection is made between it and the remote system. As the comment following this line indicates, pcANYWHERE knows this happens when it receives a data carrier detect (DCD) signal from the other computer.

The SEND CHAR command transmits a single character or control code to the remote system. In this example, you're transmitting a carriage return (ASCII character 13) to "wake up" the CIS service.

WAIT STRING is the pcANYWHERE script command that tells your PC to wait until the character string following it is received from the remote system before continuing execution. In your sample script, for example, you're looking for a Name: prompt, which is the prompt CompuServe uses to request your user ID.

A SEND LINE command sends the character string assigned to the u variable earlier in this script, representing your CompuServe user ID. Because you haven't yet entered into the sample script, this value is currently set to the dummy variable, *xxxxx,xxxx*. If you plan to use this sample script, you'll need to substitute your personal user ID for this variable before it will execute properly.

Another WAIT STRING command is used to tell pcANYWHERE to wait until CompuServe transmits its Password: prompt, which (logically enough) is the prompt used to request your password.

The final SEND LINE command is used to send the character string assigned to the p variable earlier in this script, your CompuServe password. Again, you haven't yet entered your personal password into the sample script. This value is currently set to the dummy variable, *yyy,yyyyy*. If you plan to use this sample script, you'll need to substitute the appropriate password for this variable before it will execute properly.

A general design structure

If you break this script down to its component parts, therefore, the following design framework emerges:

- Comment lines are used to define the script's purpose.
- Several LET statements identify specific information used later to sign on to CompuServe.
- A group of statements specify the appropriate configuration (terminal emulation and line settings) for CompuServe.
- A DIAL NUMBER command establishes contact with the CIS local access number.
- A series of statements actually sign you on to CompuServe and, ultimately, advance you to that service's main menu.

This sample script, although simple, reflects the kind of logical structure you should strive to incorporate into any pcANYWHERE scripts you create. Specifically:

- Include some initial comment lines to describe the script's primary function.
- Use the LET command to define any variables this script contains. (Placing variable near the beginning of your script makes them easy to find and change, if necessary.)
- Include the appropriate script commands to prepare pcANYWHERE and your modem for the current session.
- Contact the remote system, using the CALL command and pertinent sign-on procedures.
- Expand your script, as necessary, to execute the commands required to perform the specific activities you want to accomplish on this remote computer.

Common pcANYWHERE script commands

The pcANYWHERE script language includes over 130 commands you can use when designing and creating pcANYWHERE scripts. Many of these are advanced functions that you'll either never use, or won't understand how to use properly until you're much more familiar with telecommunications in general and pcANYWHERE specifically. (After all, learning everything there is to know about anything takes time.) Other script commands, however, you'll use quite often—in some cases, for virtually every script you create. I'll devote the remainder of this chapter to listing the more common pcANYWHERE script commands, and explaining their proper syntax and usage.

Note: The pcANYWHERE User's Guide contains a complete listing of the program's script commands, along with instructions on how each command is used and what it allows you to accomplish. Within the following command description, I'll use the following conventions to outline how each command and its associated parameters are entered:

- If a given command or parameter must be entered in UPPERCASE letters, that's how it will be shown here.
- Parameter variables will be shown in lowercase italics, as in the following example, where *seconds* is a variable indicating the number of seconds the script should wait:

WAIT TIME [*seconds*]

- Optional parameters will be enclosed in brackets, as in both the previous example and the following example, where the date is entered only if you don't mean today, i.e., the current date:

WAIT UNTIL *time* [*date*]

- If quotation marks (" ") are shown in an example, those quotation marks must be included whenever you actually use that command in a pcANYWHERE script, as in the following example:

WAIT STRING "Enter password:"

Where appropriate, after listing the correct command syntax for each pcANYWHERE script command shown here, I've included one or more examples of how that command might be used, along with a description of what these examples would accomplish during script execution. If you're ready, here we go.

BEEP

Sounds an alarm in the form of an audible beep. You could use this command to alert you when something important happens within your script—say, when an on-line connection is made with a remote system.

BYE

Halts execution of a script file, but leaves the current on-line session open—that is, it doesn't disconnect the line.

CD

Changes to the specified drive and/or directory. The syntax is:

CD [*drive:*]*directory path*

For example:

CD C:\\DOSFILES

This example makes the DOSFILES directory on drive C the active directory. You can use the CHDIR command to control where uploaded and downloaded files are located.

CLEAR SCREEN

Clears the entire display and positions the cursor in the upper left-hand corner of the screen. This command is useful for removing clutter from your display during an automated procedure, as you might want to do before asking for input.

COPY

Copies a DOS file from one location to another. The syntax is:

COPY *source_file destination_file*

For example:

COPY C:\AWR\READ.ME C:\DOSFILES\READ.ME

CURSOR OFF

Turns the cursor off.

CURSOR ON

Turns the cursor on.

DIAL HOST

Dials the number associated with a specific host listing. The syntax is:

DIAL HOST *"host listing"*

For example:

DIAL HOST "MY HOST"

This example dials the number recorded in MY HOST on pcANYWHERE/Remote's host listing.

DIAL NUMBER

Dials the telephone number enclosed in quotations after the command. The syntax is:

DIAL NUMBER *"number"*

For example:

DIAL NUMBER "555-5555"

This example dials the local telephone number 555-5555.

EMULATE

Sets pcANYWHERE to emulate a specified terminal type. The syntax is:

EMULATE *terminal*

For example:

```
EMULATE VT100
```

This example changes your terminal emulation from its current setting to a DEC VT100 terminal.

END

Ends the current script and restores pcANYWHERE back to where it was when this script started execution. If the current script was called from another script, pcANYWHERE returns you to the line in the original script immediately after the one calling the current script.

END TERMINAL

Ends the current script and passes control to pcANYWHERE terminal mode, regardless of what you were doing when the script started execution.

EXIT

Terminates the current script, quits pcANYWHERE, and returns you to DOS.

GOSUB

Provides an unconditional branch to a procedure associated with the specified label, and allows you to exit that procedure with a RETURN statement. The syntax is:

GOSUB *label*

For example:

```
GOSUB GETPASS
```

This example will jump to and begin executing the procedure identified with the label GETPASS. Labels are identified in a pcANYWHERE script with a colon.

GOTO

Like the GOSUB command, GOTO provides an unconditional branch to a procedure associated with the specified label. A GOTO procedure, however, won't loop back to its own location when it encounters a RETURN statement. The syntax is:

GOTO *label*

For example:

```
GOTO SIGNOFF
```

This example will jump to and begin executing the procedure identified with the label SIGNOFF (which you must write into your pcANYWHERE script). You might use this, for example, to jump to a subroutine that immediately ends an on-line session if you determine there's nothing of interest waiting for you there.

HANG

Drops a phone line connection and ends the current on-line session.

IF

Executes a series of commands or a pcANYWHERE procedure only IF the result of a conditional test is TRUE. It has the following syntax:

IF *expression* GOTO *label*
IF *expression* THEN *command*

For example:

```
IF ($RESULT) < 0) THEN TYPE STRING "Result is less than zero."
```

LET

An optional command used to assign text string or numerical values to a variable. The syntax is:

LET a = *arithmetic expression*
LET a = *"text string"*

For example:

```
LET n = "Jack Nimersheim"
```

This example assigns my name to the string variable n. I might use this LET statement to simplify signing on to a remote system that requests my full name as part of its log-on sequence.

LOAD CONFIG

Loads a hardware configuration entry. The configuration name specified with this command must already exist in your pcANYWHERE/Remote List of Hosts. The syntax is:

LOAD CONFIG *"host name"*

For example:

```
LOAD CONFIG "MY HOST"
```

This example will load the MY HOST configuration you've been using in the exercises throughout this book.

RETURN

Used to terminate a GOSUB procedure. Note: The RETURN command branches the script file back to the line immediately following the GOSUB command that initially called the subroutine.

RUN

Used to execute a DOS program. The syntax is:

RUN *command line*

Note: Some programs that normally work on your system might not load with the RUN command due to the fact that pcANYWHERE doesn't leave enough memory available for them to run properly.

SCRIPT

Executes another script, with the syntax:

SCRIPT *script name*

Note: When the second script terminates, control is passed back to the line following the one that calls it in the original script—that is, the script containing the SCRIPT command.

SEND FILE

Sends (uploads) a file to the remote system, using the currently selected protocol. It has the following syntax:

SEND FILE *protocol "filename"*

SET

Used to control various system parameters, transmission settings, and options. The syntax is:

SET *PARAMETER value*

Some conditions you can use the PARAMETER value to specify with a SET command are:

- ATTRIBUTE (use DOS screen attribute codes)
- DTR *OFF | ON*
- ECHO *OFF | ON*
- FLOW *OFF | ON*
- PARITY *EVEN | ODD | NONE | MARK | SPACE*
- PROTOCOL (select from valid pcANYWHERE protocols)
- RECORD *OFF | ON* (turns off/on session recording)

- SPEED (select from valid bps settings for your modem)
- TIMEOUT (specify timeout time, in seconds)

TERMINAL

Exits the current script file and takes you to the pcANYWHERE Terminal Mode display.

WAIT

Used to pause execution of a script file until a specified string or condition is received from the remote system. It has the following syntax:

WAIT *PARAMETER value*

Some conditions you can use the PARAMETER value to specify with a WAIT command are:

- CARRIER (wait for DCD signal)
- RECEIVE (wait for any character from the remote computer)
- SILENCE *time* (wait until the remote system sends no signals for the specified time period)
- STRING *"string"* (wait until the specified string is received from the other computer)
- TIME *seconds* (wait the specified number of seconds.)
- UNTIL *time* [*date*] (wait until the specified time or, optionally, date.)

Summary

As I mentioned earlier, there are many more commands available to you for designing and creating your pcANYWHERE scripts. All of them are listed in the pcANYWHERE *User's Guide*, along with a full explanation and examples of each. I have attempted to describe only the more common ones here, commands you'll use on a regular basis.

As I also mentioned earlier, anyone with even minimal PC experience should be able to automate virtually any on-line activity, no matter how complex that activity might be, using either pcANYWHERE's automated procedures or the new Norton pcANYWHERE script language. The examples presented in this chapter should help get you started, as you start down "automation avenue." Just remember, the more you automate, the less you have to do. pcANYWHERE provides the tools to make automation easy; the rest is up to you.

12
CHAPTER

Creating a secure environment

Despite all of the wonderful things you can accomplish with the properly equipped PC, hooking a modem up to a computer creates one major problem—security, or, more correctly, the potential for a lack thereof. As I've stated several times throughout the past eleven chapters, telecommunications opens an electronic doorway to your system.

That's fine, so long as you're familiar with everyone who takes advantage of this congeniality and drops by for a visit. But you wouldn't leave the front door to your home wide open, allowing just anyone to saunter through, would you? Admittedly, life would be a wonderful thing if you could; common sense dictates a slightly more judicious approach, however. In today's world, like it or not, such a *laissez faire* attitude is impractical, at best, both in your personal life and in your PC activities. Luckily, pcANYWHERE provides several security features that allow you to unlock the electronic door to your host system without fear of unauthorized entry.

pcANYWHERE/Host security options

The most logical place to begin a discussion of pcANYWHERE's security features is with the Security option on the pcANYWHERE/Host Preferences menu. So, get ready to switch gears again and switch over to your host computer. To access the pcANYWHERE/Host Security settings:

1. At your host computer, type CD \AWH to make AWH the active directory and press Enter.
2. Type AWHOST to start pcANYWHERE/Host and press Enter.
3. At the pcANYWHERE/Host Main Menu, highlight Preferences and press Enter.
4. When asked, type in the master password you selected for pcANYWHERE/Host and press Enter.
5. At the Host Preferences menu, highlight Security and press Enter.

pcANYWHERE will display the Host Security Preferences dialog box, shown in Fig. 12-1. You'll use this dialog box to specify those security settings you want applied globally to your pcANYWHERE/Host environment. Several of these settings are related to one another, so understanding how each works will help you establish the appropriate level of security for your host system.

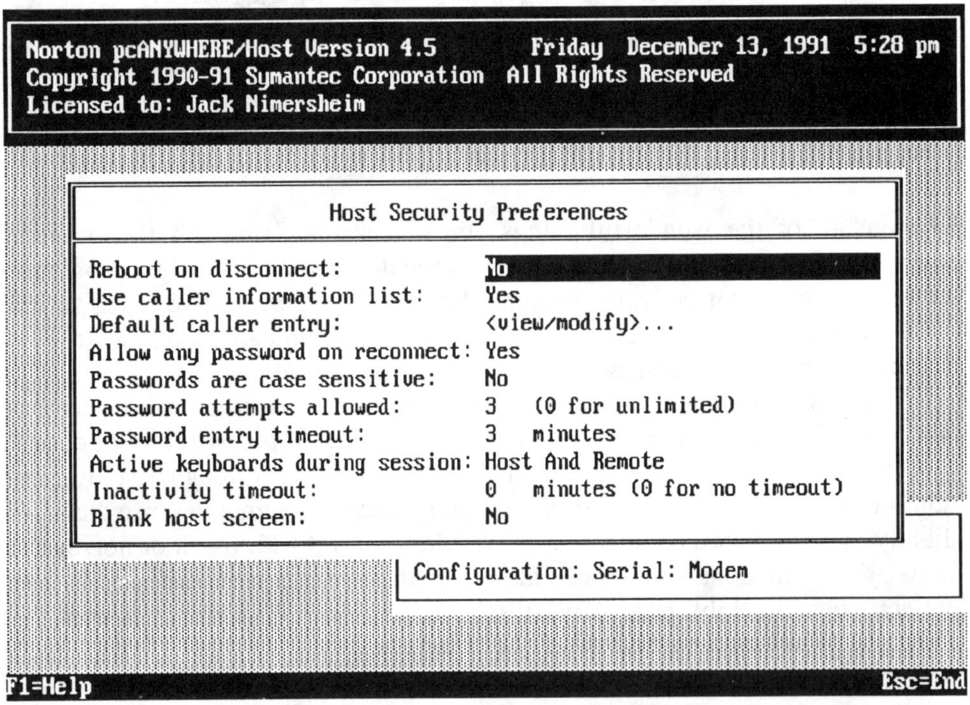

12-1 Several fields in the pcANYWHERE/Host Security Preferences dialog box are designed to prevent unauthorized callers from accessing your host system.

Note: pcANYWHERE allows you to modify selected security attributes for individual remote systems based on settings stored in their Caller Information file. The Caller Subject to Inactivity Time field in the Caller Information file, for example, can be used to override the global Inactivity Timeout setting specified in the Security Preferences dialog box for a specific remote system.

Global security options

You can specify the following settings using pcANYWHERE's Host Security Preferences dialog box:

Reboot on Disconnect The setting in this field determines what happens on your host system when the connection between the host and remote computers is terminated. This can be either a planned termination, as occurs each time a remote user ends a pcANYWHERE session, or the result of an accidental loss of signal due to problems with the telephone connection. Selecting one of the reboot options—that is, changing this field from its default No setting—ensures that each caller will be able to sign on only with his own password. (When this field is set to No, the next person calling in after an interrupted session will need to sign on using either the previous caller's password or the master password.) Also, without a reboot, an unintentional disconnect leaves the host system exactly as it was when the session was terminated, including any application programs that might have been running at the time. Consequently, selecting a reboot option also guards against one caller seeing work another caller has performed on your host system. Be aware, however, that selecting a reboot option also means you'll need to include the appropriate instructions to restart pcANYWHERE/Host in your AUTOEXEC.BAT file, if you want the system set up for additional remote sessions following reboot.

Use Caller Information List Setting this option to No causes pcANYWHERE/Host to use a default caller entry for all remote sessions. The only reason you'd do this is if you wanted to set up a generic host system, where every caller used the same password to gain access and was granted exactly the same privileges following sign on. In most cases, you'll want to leave this field set to Yes, which causes pcANYWHERE/Host to determine a caller's configuration and security privileges based on information contained in individual Caller Information files.

Default Caller Entry If you specify No in the previous field, you can use this option to display and modify the default caller entry.

Allow Any Password on Reconnect If set to Yes, callers will be able to access your host with their own passwords, even if a call is terminated unexpectedly, as outlined in the discussion under *Reboot on Disconnect*. While the Yes setting ensures that no callers will be inadvertently locked out of your system, it also means a possibility exists that one caller might see sensitive data left on the screen by the previous caller should an accidental disconnect occur.

Passwords Are Case Sensitive Changing this field to Yes adds an additional layer of security to your pcANYWHERE environment by forcing callers to enter their passwords *exactly* as they're defined in the Caller Information file, including the appropriate use of upper- and lowercase letters.

Password Attempts Allowed Limiting the number of times a caller can attempt to enter his password before pcANYWHERE/Host automatically drops a connection guards against an unauthorized caller gaining access to your system by entering a series of passwords, hoping to stumble on a valid one.

Password Entry Timeout In addition to limiting the number of times callers can attempt to enter a valid password, you can also place a time limit on how long they're allowed to spend in this effort. The time setting specified here overrides the number of password attempts indicated in the previous field.

Active Keyboards during Session The setting in this field determines whether the remote keyboard, host keyboard, or both are active during a remote session. Setting this field to Remote Only, for example, means you will be unable to access the chat feature in order to communicate with remote users during a pcANYWHERE session.

Inactivity Timeout If a value other than 0 is entered in this field, pcANYWHERE/Host automatically drops a remote connection if that number of minutes passes without keyboard or file transfer activity on either the host or remote system.

Blank Host Screen This is an extremely useful security setting, in that it keeps prying eyes from seeing any work being performed on the host system by a remote caller. Callers with Yes specified in the Allow Caller to Blank Host Screen field of their Caller Information file can override the setting in this field during a remote session.

How many and which of these settings you modify is entirely up to you, and depends largely on the level of security you feel is appropriate for your pcANYWHERE/Host environment. As a rule, the more sensitive the type of work performed on your host system is, the more concerned you should be with guarding that data against unauthorized access. Realize, also, that the settings specified in the Security Preferences dialog box represent only one layer of pcANYWHERE's built-in security features. Several fields in your individual Caller Information files complement or, in some cases, override the values entered here. I'll discuss those settings in the next section. For now, let's leave the Security Preferences dialog box and return to the pcANYWHERE/Host Main Menu. To exit Preferences and return to the Main Menu, press Esc twice.

Granting privileges to individual callers

I'm going to assume for a moment that you're setting up a fairly complex pcANYWHERE environment, one in which several callers will be accessing your host system. Furthermore, these various callers will be performing a wide range of on-line activities, with different people being responsible for different operations. In such an environment, the ability to assign varying levels of security to individual callers can be critical.

Some callers, for example, could require almost total access to your host system. Others, however, you might want to manage a little more closely, or even control completely what they are and aren't allowed to do after making a remote connection. Several fields in the Caller Information file allow you to vary specific security settings for individual callers.

To access a Caller Information file, highlight View/Modify List of Callers on the pcANYWHERE/Host Main Menu and press Enter. At the Caller Names options box, select MY REMOTE and press Enter. This displays the Caller Information dialog box shown in Fig. 12-2. (You might recall this particular Caller Information file from chapter 6, when you first set up your host environment.)

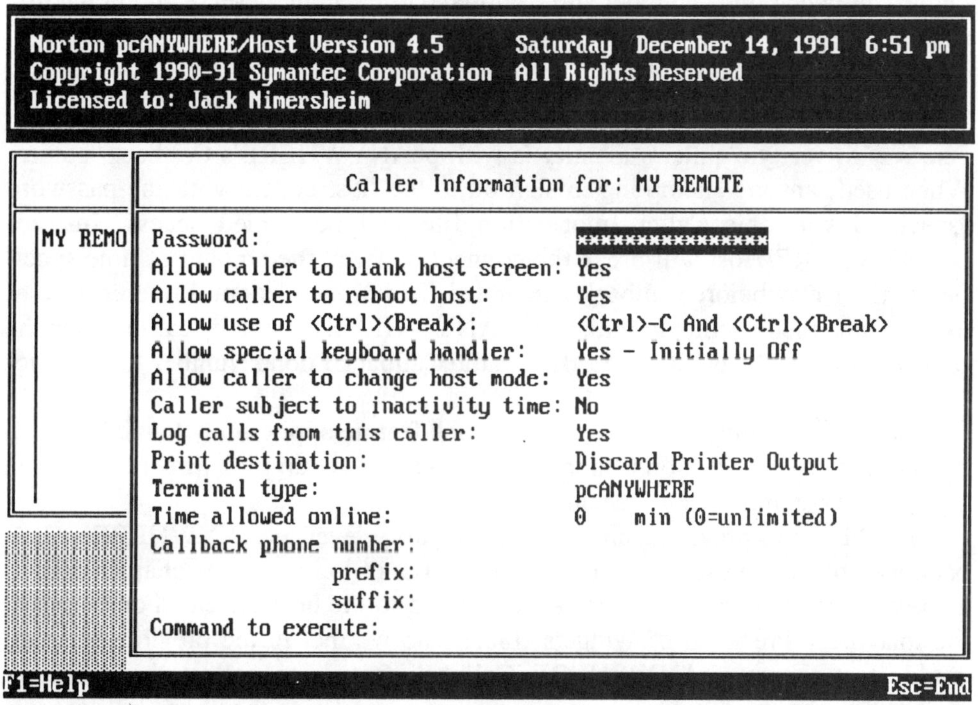

12-2 Security-related settings also exist in each Caller Information file.

Keeping in mind the various Security Preferences described in the previous section, pay particular attention to the following fields:

Allow Caller to Blank Host Screen If set to Yes, as it is in Fig. 12-2, callers signing on to your host with the password associated with this Caller Information file can override the setting specified in the Blank Host Screen field of the Security Preferences dialog box.

Allow Caller to Reboot Host If set to Yes, callers regulated by this Caller Information file will be able to manually reboot your host system by selecting the appropriate option from their remote Sessions menu.

Caller Subject to Inactivity Timeout The setting in this field overrides the Inactivity Timeout setting specified in your Security Preferences dialog box.

Time Allowed Online In addition to limiting how long a caller has to enter a valid password, you can also specify the total amount of time a caller can remain connected to your pcANYWHERE host system, following a successful sign on.

In addition to these general security settings, the Caller Information file contains two fields that give you very specific control over how a remote caller can contact your host system and what he can do after that contact is established. These two fields, Callback Phone Number and Command to Execute, deserve additional discussion.

Using pcANYWHERE's Callback feature

One way to verify a caller's identity is with pcANYWHERE's Callback feature. When used, anyone attempting to sign on to the host system with the password associated with this Caller Information file will be denied access. Instead, pcANYWHERE/Host will break the connection, wait the amount of time specified in the Delay before Callback Attempted field of the General Preferences dialog box, and then attempt to initiate a pcANYWHERE session by calling back the remote system at the number specified in the Callback Phone Number field in Fig. 12-2.

Note: As you might suspect, the Callback feature applies only to modem connections. It has no effect on Caller Information files set up for a direct connection between a host and remote system.

In addition to providing an additional security level, pcANYWHERE's Callback feature can be used to ensure that the costs of long-distance charges associated with remote connections are always billed to the host system. For example, the imaginary President of Widgets, Inc., who we met in the previous chapter, could elect to use pcANYWHERE Callback feature to collect the overnight reports from her regional sales managers, thus centralizing the phone bills for this on-line activity in the company's corporate headquarters.

Controlling an on-line session

The final field in the Caller Information dialog box, Command to Execute, is potentially pcANYWHERE's most comprehensive security feature. Used properly, it allows you to take complete control over what a remote caller can and cannot do, once he signs on to your host system.

Suppose, for example, that the only thing you want a given remote user to be able to do on your host system is download a specific file. Entering the appropriate AWSEND command in the Command to Execute field of that caller's information file would make this possible.

Note: You should change the Allow Use of Ctrl-Break setting to No, whenever you plan to use the Command to Execute field for security purposes. Doing so will prevent the remote caller from interrupting your specified routine and taking direct control of the host system.

Security on the remote system

For obvious reasons, pcANYWHERE's built-in security features are concentrated at the host end of a host/remote connection. The host system, after all, is where the greater potential for unauthorized access exists. This doesn't mean, however, that you should totally ignore security precautions when using pcANYWHERE/Remote.

Because pcANYWHERE/Remote includes an option to wait for and automatically answer incoming calls, it's also vulnerable to unauthorized access. Still, a potential intruder would need to be very determined and extremely adroit technically before he could actually take over a remote system through a pcANYWHERE connection. (The primary function of the remote computer during a pcANYWHERE session, after all, is to sacrifice its own autonomy and assume a role of subservience to the host system.)

Although a remote computer isn't totally immune to unauthorized access, it faces a much greater threat when involved in more traditional on-line activities. The source of this threat is one of the more distressing aspects of telecommunications—computer viruses.

As much as you might like to believe otherwise, some ruthless people own computers and modems. With an estimated 80 million PC users out there, there are bound to be a few rotten apples in the bunch, so to speak. Don't misunderstand; 99.99% of the people who telecommunicate are honest and trustworthy individuals. Still, there are malcontents out there whose idea of a good time is to mess up someone else's PC. One way to do this is to create and distribute destructive bits of code, called computer viruses.

Over the years, the general phrase *computer virus* has been used to describe several different kinds of potentially destructive programs: bombs, true viruses, worms, and so forth. Although each of these differs in the precise manner in which it works and what it does, they all share a common goal—to disrupt the normal operations of any computer they infect.

Some viruses are quite benign. They might simply interrupt a program you're using to display a short message and then restore the status quo, leaving things exactly as they were before the relatively minor electronic prank. A truly malevolent virus, on the other hand, can totally wipe out the contents of your hard disk in a nanosecond or two. Whether mischievous or malicious, computer viruses represent the dark side of telecommunications. Understanding the various types of viruses that exist and knowing how to combat them can help you keep your remote system healthy.

Software bombs

One extremely easy type of virus to create is the software bomb. As a rule, a software virus will invade your system disguised as a relatively innocuous program—a short piece of code that plays a brief tune, displays a cute picture, or some other equally innocuous activity. As you're distracted by a bomb's overt activities, however, it's covertly preparing to wipe out important information on your hard disk. Folks who create software bombs seem to enjoy gloating about their accomplishments; it's not unusual for a software bomb to display some sort of taunting message—like "Say bye, bye, data!" or some similar jibe—right before it completes its true mission.

Slightly more subtle, but infinitely more difficult to detect, is the so-called *logic bomb*. Unlike software bombs, which tend to entertain and then explode, thus living up to their name, logic bombs generally invade a PC embedded with a program disguising itself as a useful application or utility. The bomb itself then sits idle until it's finally "detonated" by some external event—a predetermined date, the selection of an obscure menu option, or similar activity. Once it explodes, a logic bomb quickly turns malicious, often destroying months of work in a matter of seconds. Particularly abhorrent is the fact that the people responsible for logic bombs often try to pass their vile inventions off as shareware, thus blemishing the reputation of honest individuals who rely on this distribution channel to disseminate inexpensive and useful programs.

Will the real virus please stand up?

True viruses up the ante from software and logic bombs. Mimicking its namesake, a computer virus enters your system and then starts reproducing, often attaching itself to popular commercial application programs. Once a virus has embedded itself within a program, it's virtually undetectable. Consequently, viruses are extremely difficult to identify and even harder to eliminate. A single virus entering your system might create several copies of itself, within several different programs, all within a few minutes of invading your PC. Eradicating a virus from one program, therefore, doesn't guarantee that you've eliminated it entirely from your system. One or more copies of it could remain, hidden away within other applications.

Even if a virus isn't destructive, its very presence on your PC can have an adverse effect on system performance. Each time a virus reproduces, it actually increases the size of the program file it's invaded. Over time, all this extra code can slow down disk access and, in some cases, the speed with which the application itself executes.

Worms

Metaphorically speaking, a worm is an electronic cousin of the virus. Technically, a worm is merely a neutered virus. It looks and acts just like a virus, except that worms don't reproduce. Once a worm is detected, therefore, it's relatively easy to eliminate. A single "search and destroy mission" can remove a worm from your hard disk. Undetected, a worm can still wreak havoc on your PC. But detecting and eliminating a worm is a much easier undertaking than assuring that a system is virus-free.

Fighting viruses

Now that I've cast a dark cloud over your on-line activities, let me surround it with a silver lining. The good news is that, despite their high visibility and the extensive press coverage viruses receive, the likelihood of one invading your system remains relatively small. I've been going on-line for almost a decade, and have yet to pick up a virus (knock on silicon). All commercial information services (CompuServe, Prodigy, America Online, and the like) go to great lengths to ensure that the shareware, public-domain programs, and other files they offer remain virus-free. The vast majority of local bulletin board system (BBS) operators are equally conscientious.

Complementing these efforts with a few defensive measures of your own will virtually guarantee that you need never suffer the slings and arrows of a viral attack. Avoiding a virus infection, after all, is one area where, to quote the old cliche, "an ounce of prevention is worth a pound of cure."

First and foremost, back up your hard disk regularly. That way, even if a virus does manage to invade your system, you'll be able to recover any files it infects. As a second line of defense, consider adding an antivirus program to your PC's software arsenal. Such programs, sometimes called *vaccines* (appropriately enough), can be grouped into two categories: detection software and prevention programs. Stated simply, detection software is designed to identify viruses that have already invaded your system and then take the appropriate steps to eliminate them. A prevention program, on the other hand, acts like an electronic sentry, guarding your system against initial infection.

Symantec, the company that manufactures pcANYWHERE, also offers an antivirus program called Norton Anti-Virus. Norton Anti-Virus contains both a detection and a prevention module. The first, a 24K terminate-and-stay-resident

(TSR) program is designed to intercept potential viruses entering your system and alert you to their presence before your PC or disk can be affected. The second Norton Anti-Virus module, an executable program called Viral Clinic, examines your disks for any viruses it might already contain.

If Viral Clinic discovers a virus, it will present you with an appropriate list of options and suggestions on ways to eliminate the infection. In some cases, Viral Clinic can actually remove a virus and then restore any programs it infiltrated to pristine condition. If this isn't possible, you can always instruct Viral Clinic to completely erase the questionable file from your disk—a process not unlike what happens when a surgeon excises a cancer tumor from the human body.

As I stated earlier, my goal here isn't to alarm you overly much. You could use pcANYWHERE/Remote every day for years and never encounter a bomb, a virus, a worm, or any other type of potentially destructive code. Then again, you could run pcANYWHERE/Remote tomorrow and download a seemingly useful utility from your favorite on-line service, only to discover that your system has been invaded. Taking precautionary measures to prevent this from happening doesn't imply paranoia. It merely represents a cautious attitude toward a potentially destructive PC phenomenon.

Note: While I'm on the subject of viruses, let me point out that the most common way they infiltrate a computer isn't over a modem connection at all. Far more prevalent is the situation where a virus invades a system from a floppy disk. Specifically, it's not unusual for viruses to be hidden away in unauthorized (or pirate) copies of a commercial application program. Consequently, accepting a pirate program isn't just illegal, it's also not very smart. You might think you're getting a bargain by picking up a $400 or $500 program for next to nothing. The true cost of your actions, however, could be the loss of priceless data. Consider this the next time someone offers you a deal that "seems too good to be true." It probably is.

Summary

Unfortunately, security is something that must be considered whenever you open an electronic doorway to your PC. Fortunately, as you've seen in this chapter, pcANYWHERE provides a wide range of security-related features. Taking full advantage of these security features, and supplementing your efforts with some practical measures to protect your host and remote computers against a potentially destructive virus infection, will almost guarantee their safety.

13
CHAPTER

The pcANYWHERE utilities

Utilities. Say this word to some people and their eyes will light up as they contemplate all the nifty things utilities allow them to accomplish, in an ongoing effort to wring the *n*th degree of flexibility and functionality out of their favorite applications. Mention utilities to the vast majority of users, however, and they'll cringe. Why? Because, unfortunately, the utilities that accompany many PC programs are confusing and convoluted program modules that could be useful if only you could figure out how to use them.

Both pcANYWHERE/Host and pcANYWHERE/Remote include a number of utilities. That's the bad news. The good news is that the pcANYWHERE utilities really are easy to use. As is true with virtually every other pcANYWHERE feature, its utilities rely on interactive menus, option lists, and dialog boxes to demystify potentially confusing and often technical procedures. Consequently, even relative newcomers to telecommunications will have no trouble "tweaking" their pcANYWHERE environment to peak performance.

Accessing the pcANYWHERE utilities

The Main Menu in both pcANYWHERE/Host and pcANYWHERE/Remote contains a Utilities option. As you would expect, you use this option to access the various utilities available on either your host or remote system.

To display the Utilities menu, highlight Utilities and press Enter. If you're working with pcANYWHERE/Remote, selecting Utilities will display the menu shown in Fig. 13-1. Figure 13-2 contains the corresponding menu from pcANYWHERE/Host. The various options on these menus fall into three general categories:

- system diagnostics
- activity reporting
- customization tools

The individual utilities in the first category are identical to one another on both the host and remote systems. Activity reports and customization options will vary, depending on whether you're working with pcANYWHERE/Host or pcANYWHERE/Remote.

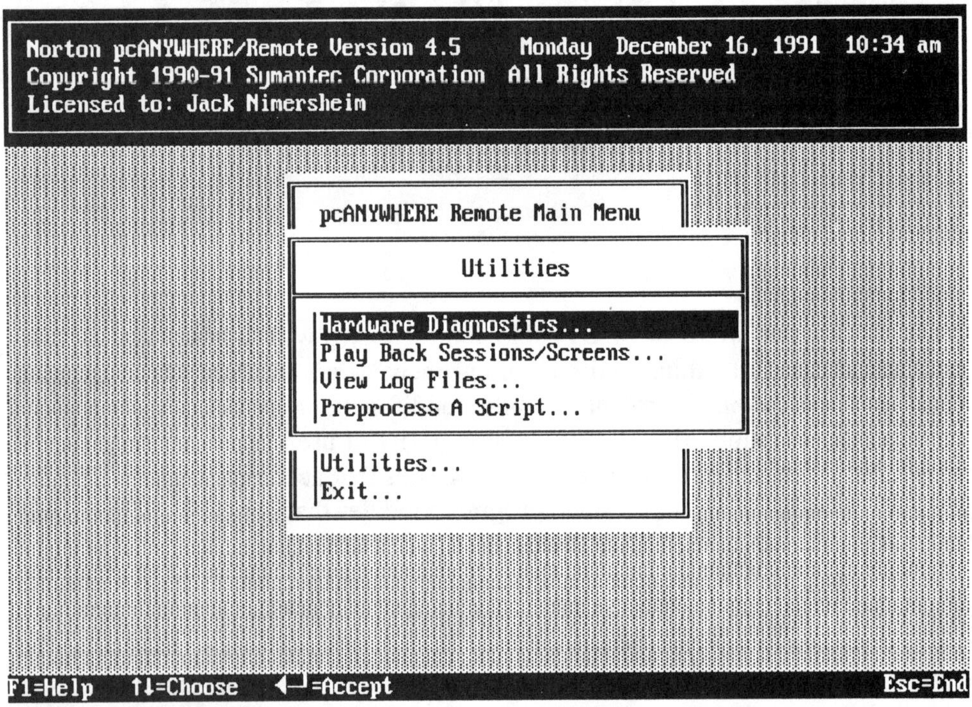

13-1 The pcANYWHERE/Remote Utilities Menu.

pcANYWHERE's hardware diagnostics

We'll begin by looking at pcANYWHERE's diagnostic utilities. As mentioned in the previous section, these utilities are the same in both pcANYWHERE/Host and pcANYWHERE/Remote. To access pcANYWHERE's diagnostic utilities, highlight Hardware Diagnostics and press Enter.

Selecting Hardware Diagnostics on either the host or remote system displays

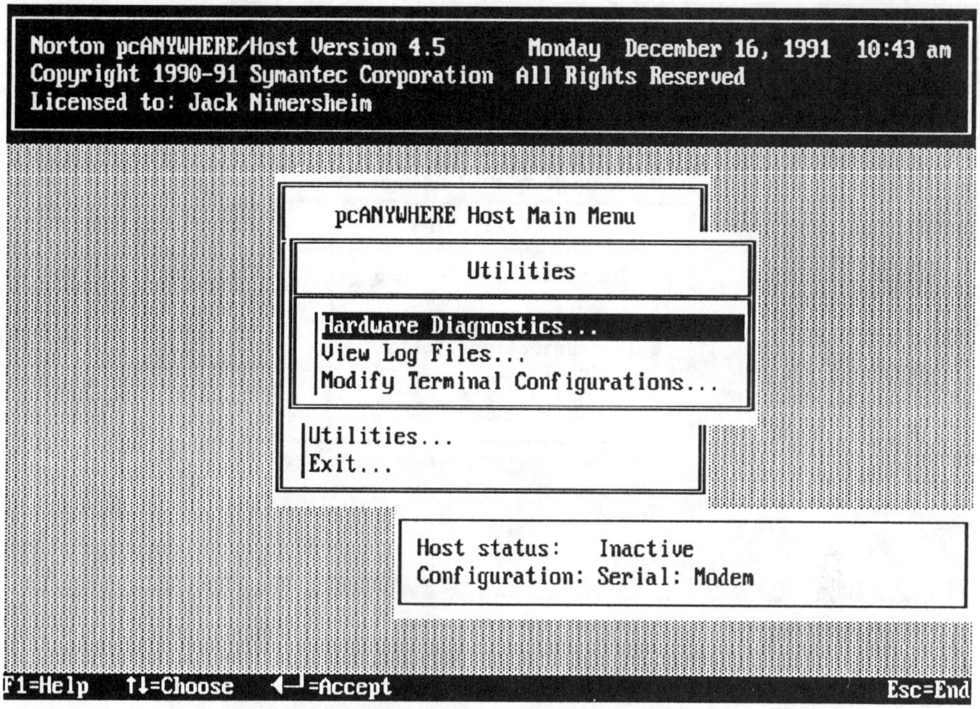

13-2 The pcANYWHERE/Host Utilities Menu.

the Diagnostics Main Menu, shown in Fig. 13-3. The various options available on this menu allow you analyze and, when appropriate, test the hardware and software installed on your host and remote systems. Used properly, pcANYWHERE's hardware diagnostics can help you identify the source of any problems you might encounter when setting up your remote-computing environment. Options available on the Diagnostics Main Menu include:

Overview Use this option to display basic information about your PC. In addition to listing the specific hardware available on your system—display type, communications ports, keyboard, etc.—the Overview option lets you check the current settings in your DOS environment, comparable to running the DOS SET command, as well as view the contents of your AUTOEXEC.BAT and CONFIG.SYS files. Overview's primary function is to help you track down potential conflicts between pcANYWHERE and your various system components.

Communications Tests Select this option to run a series of hardware tests based on the settings specified in a particular pcANYWHERE hardware configuration. Items checked include the status of your COM port and, when applicable, whether or not pcANYWHERE is able to successfully establish communications with your modem using the parameters specified in the selected hardware config-

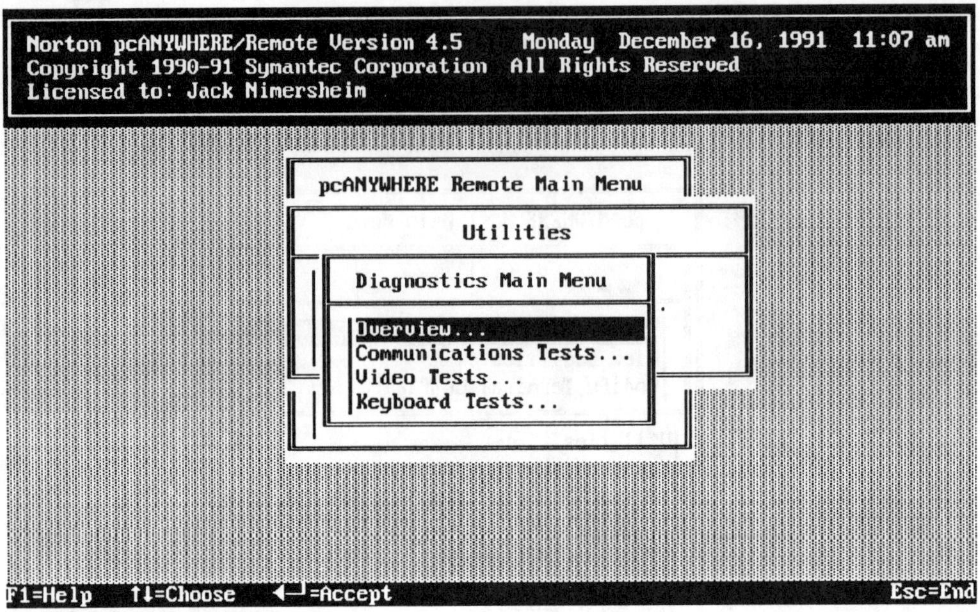

13-3 The Diagnostics Main Menu.

uration. (Configurations in which the modem field is set to None—that is, configurations designed to support a null-modem cable connection between the host and remote systems—won't include the option to test modem communications.)

Video Tests Selecting this option will advance you to a series of tests designed to check whether pcANYWHERE has any trouble communicating with your display adapter and monitor.

Keyboard Tests Use this option to access a series of tests designed to verify that pcANYWHERE is able to communicate successfully with your keyboard.

The majority of these diagnostic tests are designed to be used under the supervision of Symantec's technical support staff, should you ever need the company's assistance to resolve conflicts that surface when you're using pcANYWHERE. More often than not, however, running the Communications Tests will help you eliminate problems relating to a specific hardware configuration yourself without having to contact technical support.

To test the settings in a hardware configuration, highlight Communications Tests and press Enter. At the subsequent message screen, press Enter.

Performing communications tests

Next, pcANYWHERE displays a listing of any hardware configurations you've created on your host or remote system. To select a specific configuration, highlight the name of the hardware configuration you want to test and press Enter.

Anyone who specified a master password for their host or remote system will be asked to enter it. On systems not non-password protected, pcANYWHERE will automatically advance you to a Configuration dialog box, as shown in Fig. 13-4. After verifying that the configuration you selected is indeed the one you want to test, press Esc.

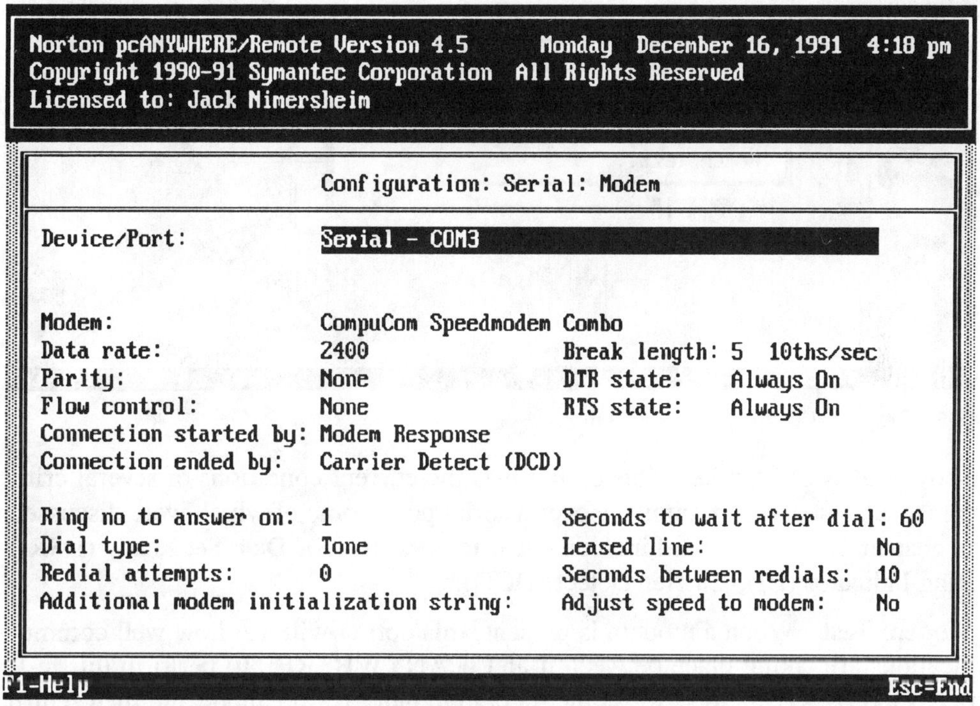

13-4 pcANYWHERE performs communication tests based on specific hardware configurations.

Note: If your pcANYWHERE host or remote system is password protected, the master password isn't required to actually perform the Communications Tests. On such systems, however, you won't be allowed to access and modify a Hardware Configuration file during testing unless you enter the master password correctly.

You'll now be presented with an option box listing the available pcANYWHERE Communications Tests. Figure 13-5 shows the options that appear for a hardware configuration in which a modem is specified. (As I indicated previously, if the configuration you're testing is designed for a null-modem connection, the Modem Test option would not show up in the Communications Tests option box.) Options in this figure include:

Check Serial Port This option tests to see whether the serial port specified in the current hardware configuration does indeed exist.

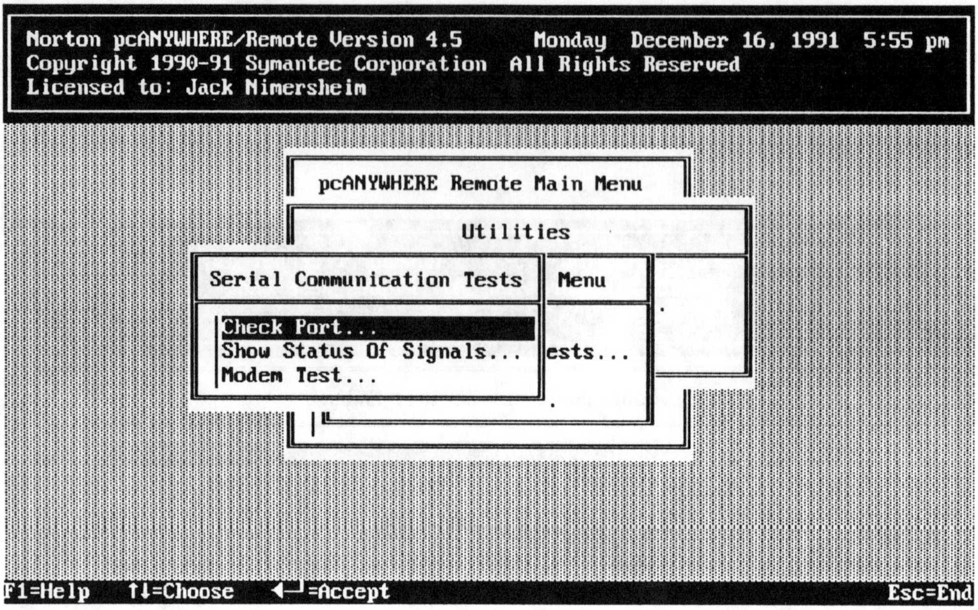

13-5 The Serial Communications Tests menu.

Show Status of Signals This option lists the current conditions of several critical signals used to communicate over a serial port, some of which were discussed in chapter 3. Signals tested include Clear to Send (CTS), Data Set Ready (DSR), Ring Indicator (RI), Carrier Detect (DCD).

Modem Test When a modem is present, this option will test how well communications are taking place between it and pcANYWHERE. To perform this test, pcANYWHERE attempts to set the modem to auto-answer mode and then return it to its initial, non-answer state. (Again, chapter 3 contains an explanation of how auto-answer mode works.) Because this procedure entails a modem acknowledging the successful receipt of a signal from pcANYWHERE, thus forcing two-way communication, it represents a legitimate—and relatively easy—way to verify whether the modem and pcANYWHERE are indeed "talking" to one another.

The various hardware diagnostics are designed strictly to help you gather information about your hardware and software. None of the tests modifies your hardware or software in any way. Consequently, experimenting with them cannot have any adverse effect on your pcANYWHERE environment. So go ahead and experiment. If nothing else, observing what happens when you perform specific pcANYWHERE diagnostic tests will be an educational experience.

When you feel educated enough, press Esc until you've returned to the Utilities menu. Now that you're back where you started, a little wiser for your journey, let's examine some of the different ways in which pcANYWHERE can help you keep track of your on-line activities.

Viewing log files

Both pcANYWHERE/Host and pcANYWHERE/Remote can be set up to maintain a record of incoming and outgoing calls to your host and remote systems. If activated, these records—called *log files*—will automatically track the time, date, and duration of all calls placed to and from a host and remote computer. Individual log entries contain additional information about the other system involved in a remote session. Stated simply, pcANYWHERE's log files are the electronic equivalent of the register books many businesses use to keep a written record of people entering and leaving their offices. Potential uses for the pcANYWHERE log files include:

- maintaining billing records
- providing an audit trail for security purposes
- project management
- efficiency tracking

As you'll see when you start examining log files, pcANYWHERE provides a great deal of flexibility in relation to what information a given report contains and how that information is formatted. So let's throw a log file on the fire, so to speak, and take a look at some of the ways pcANYWHERE lets you review records of your on-line activities.

Note: For information contained in log files to be accurate, pcANYWHERE must be told to maintain log records on your host and remote systems. This entails specifying Yes in the Log Calls from This Caller and Log Session fields of your individual Caller Information files or List of Hosts entries, respectively.

Accessing information stored in your log files

To request log information on either the host or remote computer, highlight View Log Files on the Utilities menu and press Enter. pcANYWHERE will display the Log Processing Options menu, shown in Fig. 13-6. Use this menu to manage information stored in your various log files. Available options include:

Activity Log Select this option to view log information in chronological order.

Billing Detail Select this option to view a detailed account of on-line activity, including time and purpose of each call, sorted by the name of the computer involved in each session.

Billing Summary Select this option to view a report containing only the total number of calls placed by and time spent on-line with each user. Like the billing detail report, the summary report is sorted by computer name.

Delete Old Log Data This is a maintenance function, used to remove superfluous or outdated information from a log file.

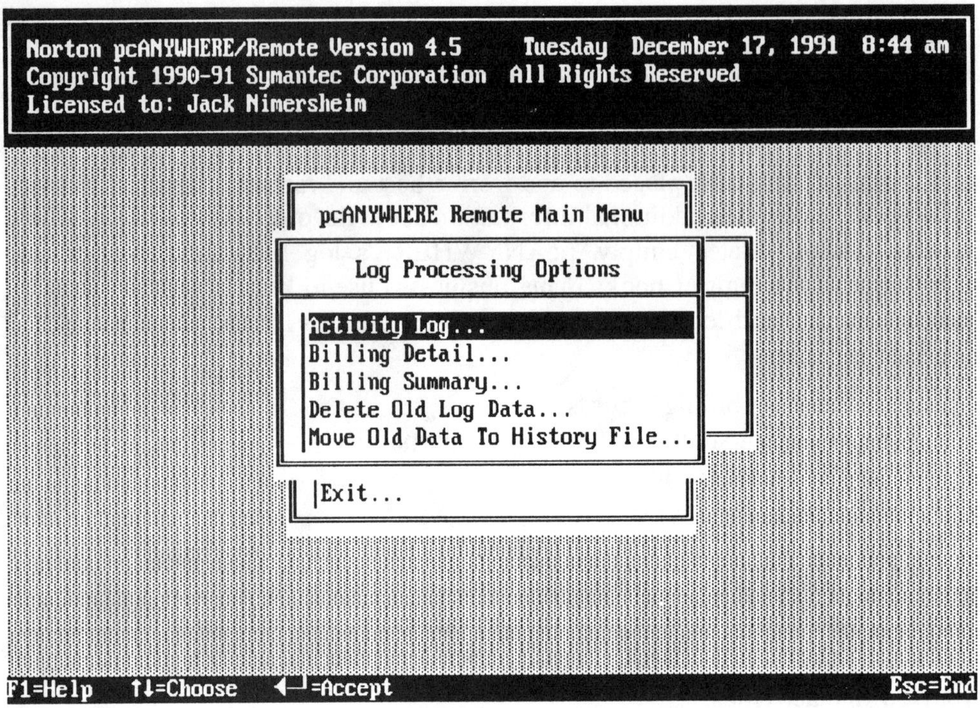

13-6 The Log Processing Options menu.

Move Old Data to History File This is another maintenance function. Rather than deleting records, however, it allows you to transfer selected records from your main log file to a second file, called a *history file*. Select this option when you want to maintain an archival copy of on-line activities that no longer need to be kept in the active log file.

Viewing your activity log

As mentioned above, an activity log contains information about your on-line activities sorted in chronological order. To generate an activity log report, highlight Activity Log and press Enter. pcANYWHERE will display the Parameters for Activity Report dialog box, shown in Fig. 13-7. Items specified in this dialog box include:

Log File to Process Use this field to identify the activity log file containing the information you want included in this report. Options include current or history files on either the host or remote system.

Start Date Use this field to specify a start date for this report, that is, the earliest date for which you want log records included in the current report. The default

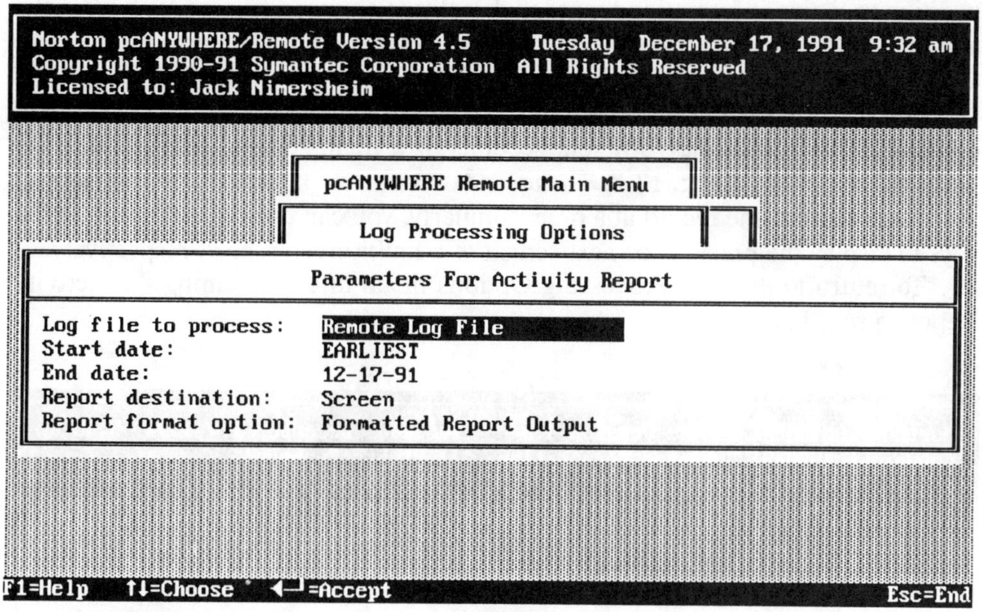

13-7 You can specify the information you want included in an activity report.

value for this field is EARLIEST, which includes all log entries from the first contact recorded in the selected file.

End Date Use this field to specify the last date for which you want log records included in the current report. The default value for this field is the current system date.

Report Destination Use this field to identify where you want the current report sent. By default, pcANYWHERE displays reports on your PC's monitor. Other options include sending the activity log report to a disk file or printing it out on either an 80- or 132-column printer.

Report Format Option Use this field to specify how pcANYWHERE should format an activity log report sent to a disk file or printer. (Activity logs sent to the screen are always displayed using the default Formatted Report Output parameters.)

For this initial report, just accept all the default values. That is, you'll request that a complete report of all log activities be displayed on the screen. To accept the default activity report parameters, press Esc. When pcANYWHERE asks you to verify that you want to process the current report, highlight Yes and press Enter.

A few seconds later, your screen should resemble Fig. 13-8, which contains an activity report generated on my system, based on the default parameters listed above. (Of course, the specific information included in your report—dates, times, descriptions, and the like—will differ from what is shown here.) Notice that, in addition to listing information recorded directly in your log file, pcANYWHERE automatically calculates the time spent on-line based on the contents of the Time On and Time Off fields for each record.

You might want to take a moment to familiarize yourself with the keyboard commands used to display different sections of a multi-screen report, which are listed on the bottom of Fig. 13-8. The up and down arrow keys, for example, let you scan large reports one record at a time. Similarly, you can use the PgUp and PgDn keys to display the previous or next screen of a multiscreen report, respectively.

To return to the Log Processing Options menu after examining the Activity Report, press Esc.

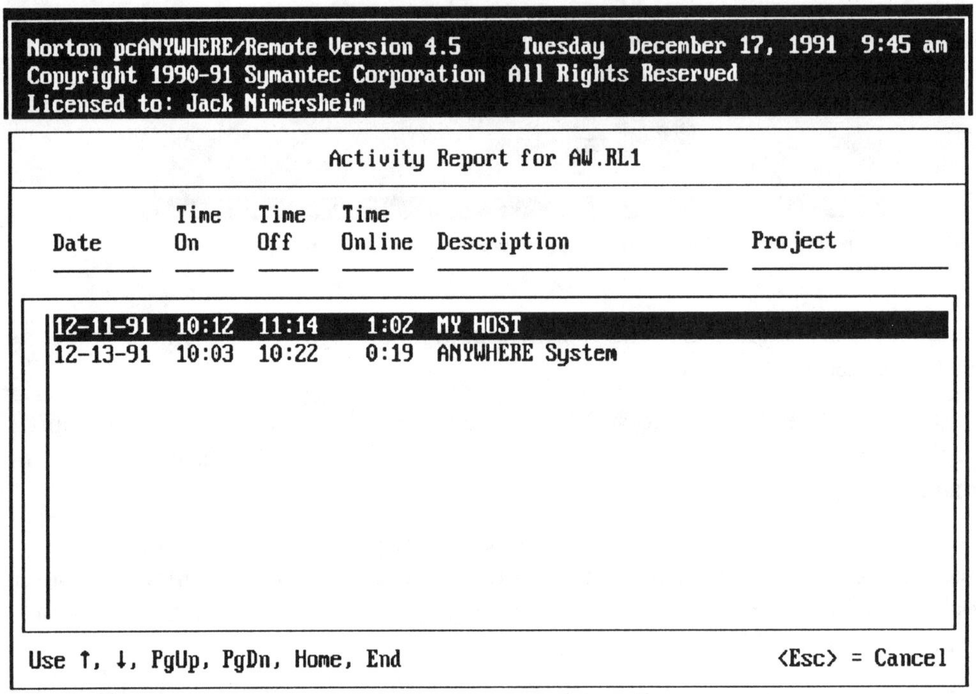

13-8 Information in an activity report is sorted chronologically.

Managing billing records

The activity log provides a comprehensive overview of all your on-line activities. Even more useful, however, is pcANYWHERE's ability to manage information that can help you organize and, in certain circumstances, generate billing records.

To examine this pcANYWHERE feature, highlight the Billing Detail option on the Log Processing Options menu and press Enter. As happened when you chose Activity Log, pcANYWHERE will display a Parameters dialog box (see Fig. 13-7). Once again, accept the default parameters by pressing Esc. When pcANY-

WHERE asks you to verify that you want to process the current report, highlight Yes and press Enter.

Figure 13-9 contains the billing detail report generated on my system using the default parameters. This new report might look similar to the activity log shown in Fig. 13-8, with one critical difference. Specifically, the individual lines in the billing detail report are sorted by the host system called, instead of being listed in chronological order as they are in an activity log. (If you generated your billing report using pcANYWHERE/Host, the individual records will be sorted by the name of each remote computer that has called the host system.) This arrangement makes it much easier to extract information for a specific customer, which can then be used to generate individual billing statements.

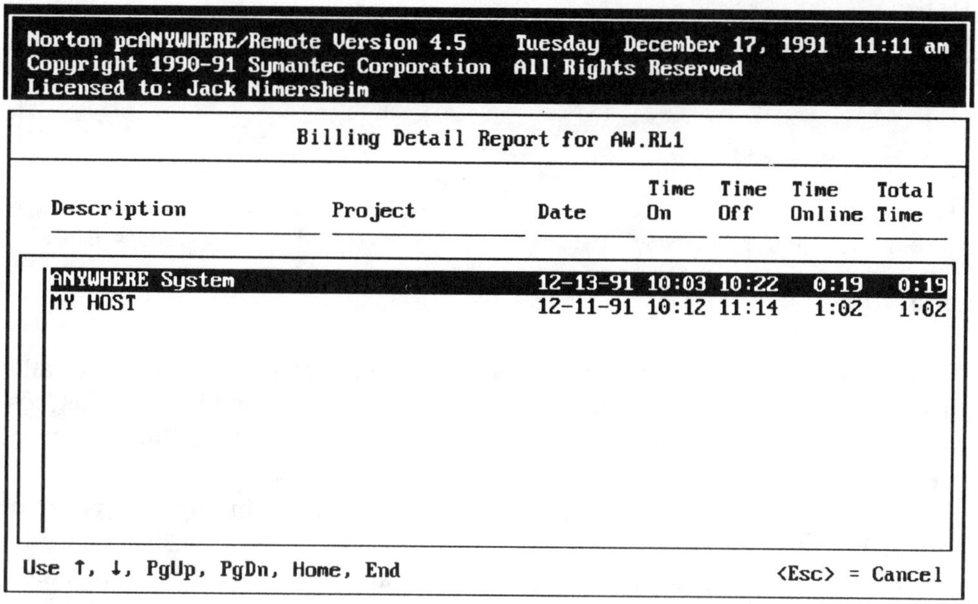

13-9 A Billing Detail report.

You say you don't need this much detail for your billing statements? Okay, then, try the following experiment:

1. Press Esc to remove the billing detail report from your display.
2. Highlight the Billing Summary option on the Log Processing Options menu and press Enter.
3. Press Esc to accept the default report parameters.
4. When pcANYWHERE asks you to verify that you want to process the current report, highlight Yes and press Enter.

Your screen should now resemble Fig. 13-10, which shows a sample billing summary report based on the contents of my pcANYWHERE log files. In this report,

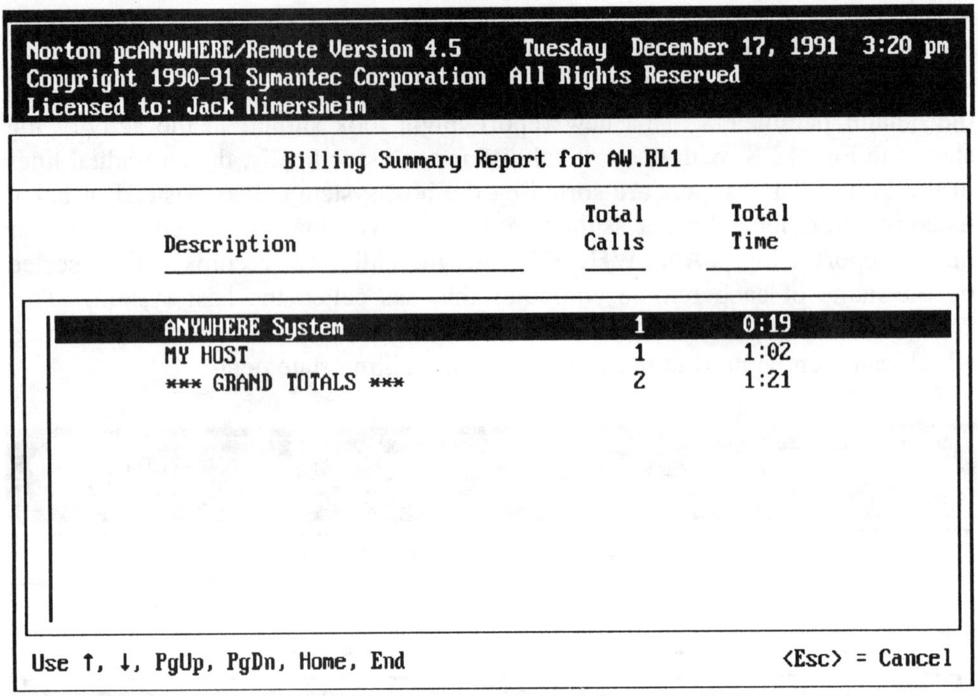

13-10 A Billing Summary report.

a single line item is generated for each customer—defined here as a separate caller or host system. Each line item lists the number of calls completed to that customer, along with total time spent on-line. After reviewing the billing summary, press Esc to return to the Utility menu.

As these simple examples illustrate, pcANYWHERE's billing reports can be extremely useful in managing your billing records. If your business is on a monthly billing cycle, for instance, entering the appropriate start and end dates at the Billing Detail or Billing Summary Parameters dialog box would provide all the information required to calculate that portion of a customer's invoice, relating to amount of time spent on-line with that customer during a given month.

But if you're going to take this first step, why not go all the way? For example, why not format your report as either a fixed-length or comma-separated data file and send it to a disk file, rather than the screen? Once this file exists, the log records or billing information it contains can be transferred easily to a spreadsheet or database manager for additional processing.

Playing back sessions and screens

Remember back in chapter 9 when you used the Save Screen option of the Sessions menu to capture some work you performed on your host system? or how about in chapter 10, when you recorded a portion of a call to MCI Mail? If you've been wondering what you were going to do with this material, wonder no longer. You're going to use it now to demonstrate another utility available on the pcANYWHERE/Remote menu: the Play Back Sessions/Screens option.

Note: Because captured screens and sessions are stored on the remote system only, Play Back Sessions/Screens isn't available in pcANYWHERE/Host. If you've been working on a host system, therefore, you'll need to switch to your remote computer if you want to perform the following exercises.

To understand how Play Back Sessions/Screens works, consider two familiar objects: a camera and a video camcorder. Back in chapter 9, you used the Save Screen option on the Sessions menu to take an electronic snapshot of the host display, in much the same way you would take pictures with an ordinary camera. Playing back this screen is comparable to looking at your pictures after they come back from the developing lab. Similarly, in chapter 10, you selected Recording On from the Terminal options menu to capture and save a portion of an MCI Mail session to a disk file—a process not unlike using a video camcorder to tape a wedding, birthday party, or some other special event. Selecting the Play Back Sessions/Screens option is similar to taking the final tape and viewing it on a VCR. So let's view your handiwork, shall we? To view a saved screen:

1. Highlight Play Back Sessions/Screens and press Enter.
2. At the Playback Options menu, highlight Play Back Saved Screens and press Enter.
3. When pcANYWHERE asks you to select the path containing the desired screen file, press Enter to accept AWR, the default directory.

pcANYWHERE will display a listing of any screen files it finds in the specified directory, as shown in Fig. 13-11. To view a particular screen, highlight its name (in this case, BUDGET.SCN), and press Enter. This will advance you to a second Playback Options menu, shown in Fig. 13-12, containing the Begin Screen Playback and Set Starting Screen Number options.

Permit me, if you will, to expand slightly on my earlier analogy. Specifically, let's take the individual pictures to which I compared captured screens and convert them into a series of slides. Using this new metaphor, a file such as BUDGET.SCN file will resemble a slide carousel, in that it can contain multiple screen shots. pcANYWHERE uses the Start Screen Playback message section of the Playback Options screen to indicate the total number of captured screens in the current file. Figure 13-12, for example, contains the following message:

```
There are 1 screens contained in:
BUDGET.SCN
```

Use the Set Starting Screen Number option to indicate where in a multiple-screen file you want to start viewing your captured screens. This process is similar to

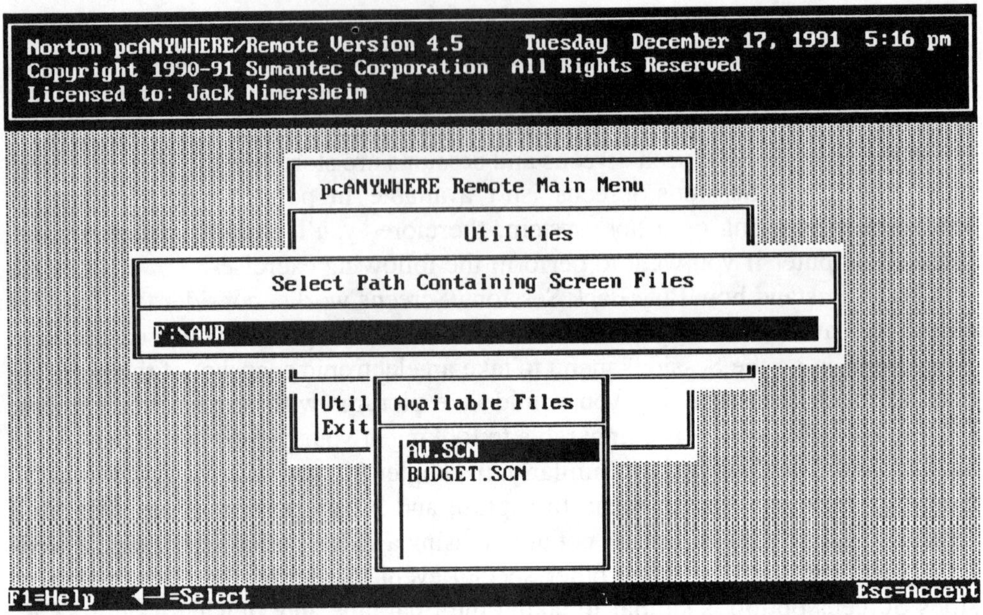

13-11 You tell pcANYWHERE which captured screen you want to see.

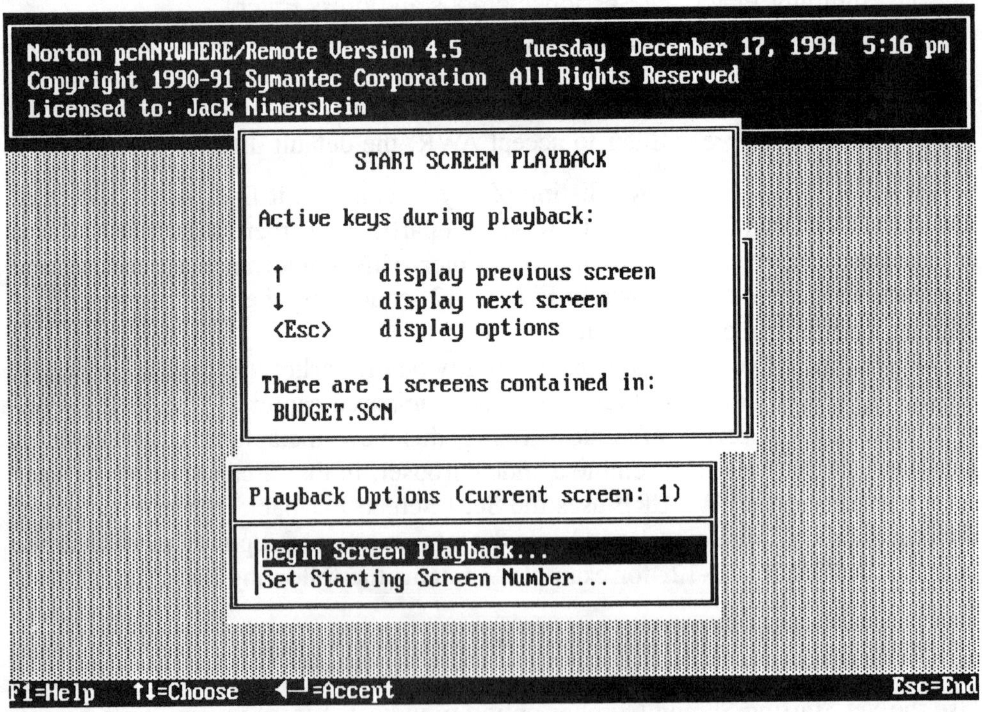

13-12 pcANYWHERE will inform you how many screens are in the selected file.

choosing the spot in a carousel at which to begin a slide show. After specifying a starting screen number, if desired, highlight Begin Screen Playback and press Enter.

This will begin your electronic "slide show." On my system, for example, selecting the Begin Screen Playback option causes pcANYWHERE to display the screen shown in Fig. 13-13, the sample Lotus 1-2-3 spreadsheet I captured during a remote session way back in chapter 9. (Check out Figs. 9-2 and 9-3 if you don't believe me.)

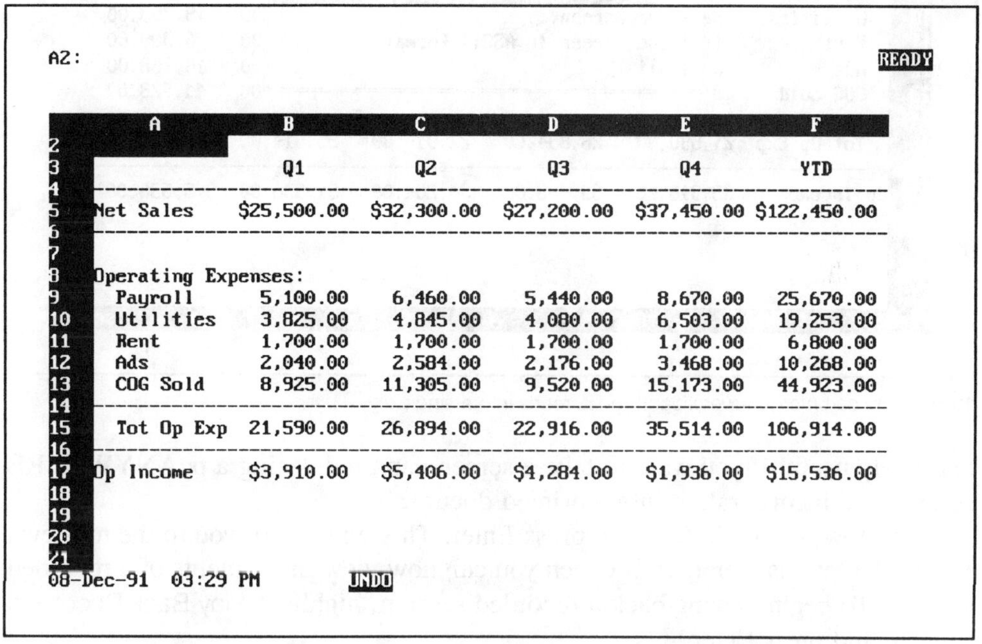

A2: READY

	A	B	C	D	E	F
2						
3		Q1	Q2	Q3	Q4	YTD
4						
5	Net Sales	$25,500.00	$32,300.00	$27,200.00	$37,450.00	$122,450.00
6						
7						
8	Operating Expenses:					
9	Payroll	5,100.00	6,460.00	5,440.00	8,670.00	25,670.00
10	Utilities	3,825.00	4,845.00	4,080.00	6,503.00	19,253.00
11	Rent	1,700.00	1,700.00	1,700.00	1,700.00	6,800.00
12	Ads	2,040.00	2,584.00	2,176.00	3,468.00	10,268.00
13	COG Sold	8,925.00	11,305.00	9,520.00	15,173.00	44,923.00
14						
15	Tot Op Exp	21,590.00	26,894.00	22,916.00	35,514.00	106,914.00
16						
17	Op Income	$3,910.00	$5,406.00	$4,284.00	$1,936.00	$15,536.00
18						
19						
20						
21						

08-Dec-91 03:29 PM UNDO

13-13 Captured screens appear exactly as they did during your remote computing session.

If viewing captured screens were all you could do with the Play Back Saved Screens option, however, it would be an entertaining but only marginally useful feature. Luckily, you can do much more. To see what I mean, press Esc. pcANYWHERE responds by displaying the Playback Options menu show in Fig. 13-14. This new menu is similar to the one in Fig. 13-12, but contains two additional items, Save Screen and Save Screen in ASCII Format.

The first option lets you combine selected screens from one or more files to create a new file, a process not unlike taking slides from several carousels and combining them into a single slide show. The second option saves the current screen to a new file in ASCII format—that is, a standard text file—without preserving colors, graphics characters, and the like. The ASCII option can be

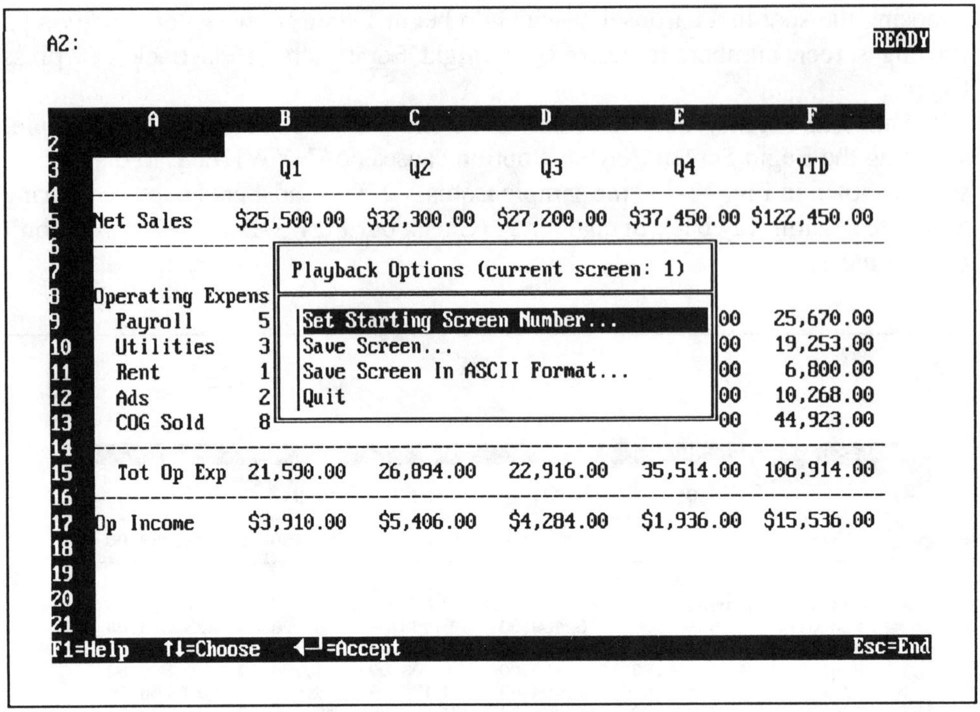

13-14 You can merge several captured screens into a single file.

extremely useful if you want to take a screen captured during a pcANYWHERE session and incorporate it into a printed document.

For now, highlight Quit and press Enter. This will return you to the top-level Playback Options menu, with which you can now view the contents of a recorded session. To begin playing back a recorded session, highlight Play Back Recorded Session and press Enter.

Once again, pcANYWHERE will walk you through the steps required to identify the recorded session you want to view, which includes specifying the path in which this session is stored and then selecting the appropriate session file. After you've provided all the necessary information, pcANYWHERE will display the Playback Options menu shown in Fig. 13-15. Notice that, when you select the Play Back Recorded Session option, pcANYWHERE will display the following options for recapturing portions of a session before actual playback begins—Start Recording to Another File and Save Session to a Raw Binary File.

The first option allows you to "cut and paste" portions of different sessions into a single file. If you've ever seen someone edit a video tape or splice a movie, you'll understand the basic concept behind how this works. The second option saves the actual data that was received by pcANYWHERE/Remote during recording, but strips away the special formatting commands the playback utility relies on to read and display captured sessions.

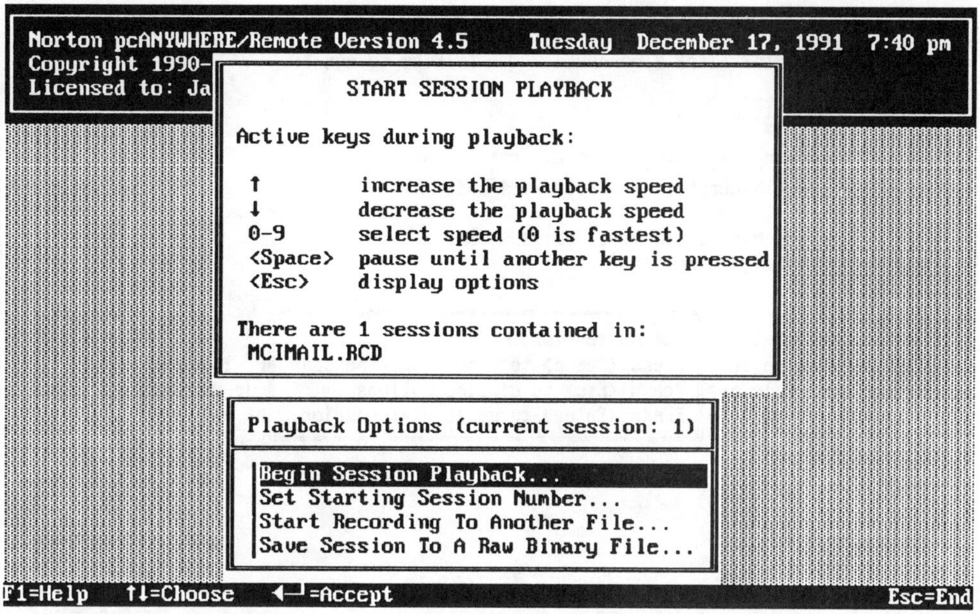

13-15 During a session playback, you select what you want to do with the session before playback begins.

Figure 13-16 shows the results of playing back the MCI Mail session that was recorded in chapter 10. Notice that one of the options presented to you when a session ends is to press the spacebar. This clears the message box in Fig. 13-16 so you can see the screen underneath. If you'll recall, the Recording On option was selected to make certain that the message waiting on MCI was delivered. As Fig. 13-16 illustrates, it worked.

To end the session playback, press Esc to redisplay the Playback Options menu. Highlight Quit and press Enter. Press Esc to exit the Play Back Sessions/ Screens option and return to the Utilities menu.

As these two examples illustrate, the ability to play back captured screens and recorded sessions can help you organize and manage information gathered on-line. Consequently, the Play Back Sessions/Screens option is a very powerful and useful feature. But you're not finished yet. Believe it or not, there are still two more pcANYWHERE utilities to discuss.

Processing scripts

In chapter 11 you analyzed a sample pcANYWHERE script, CIS.SCR, which I created with an ASCII text editor. Creating that script completed only half the job of automating my CompuServe sign-on procedures, however. Before I can actually run CIS.SCR, it needs to be converted from a standard text file into pcANYWHERE's executable SCX file format.

```
  1  Dec 10 19:47  Jack Nimersheim       Talking to myself          325

Command: READ 1

Date:      Tue Dec 10, 1991  7:47 pm  EST
From:      Jack Nimersheim / MCI ID: 494-1341

TO:      * Jack Nimersheim / MCI ID: 494-1341
Subject:  Talking to myself

Jack,
                    End of session:   1
They say you're in t Press <Space> to remove this message f. I wonder
if the same rule app Press <Esc> to display options       our own MCI
mail box. In my defe Press <Enter> to go to next session  he only way
I could think of to                                       E-mail.

Take care...um...Jack

Command: quit

Signing off from MCI Mail.
Call clearing requested by remote source.
,▬£τ/Kòσ~╖S
F1=Help   ↑↓=Choose   ◄┘=Accept                                   Esc=End
```

13-16 My sample session captured a message sent to me over MCI Mail.

To convert a script file into SCX format, highlight Preprocess a Script and press Enter. When pcANYWHERE asks you to select the path containing the desired screen file, press Enter to accept AWR, the default directory. pcANYWHERE will display a listing of any SCR (script text files) it finds in the specified directory.

To begin converting a file, highlight the name of the SCR file you want to convert and press Enter. As pcANYWHERE processes your SCR file into an executable script, each line it contains is tested for possible problems—unrecognized commands, syntax errors, and so forth. If everything works as expected, pcANYWHERE will complete the conversion process and display a message announcing its success.

To remove this message and return to the Utilities menu, press Esc. If problems do exist, the conversion process will be canceled and you'll see a message similar to the one in Fig. 13-17. As this message indicates, pcANYWHERE automatically saves a record of those errors it detects in a special file that has the same name as your original SCR file, but is assigned an extension of ERR. After examining this file and correcting any errors it contains, you should be able to reprocess your SCR file successfully.

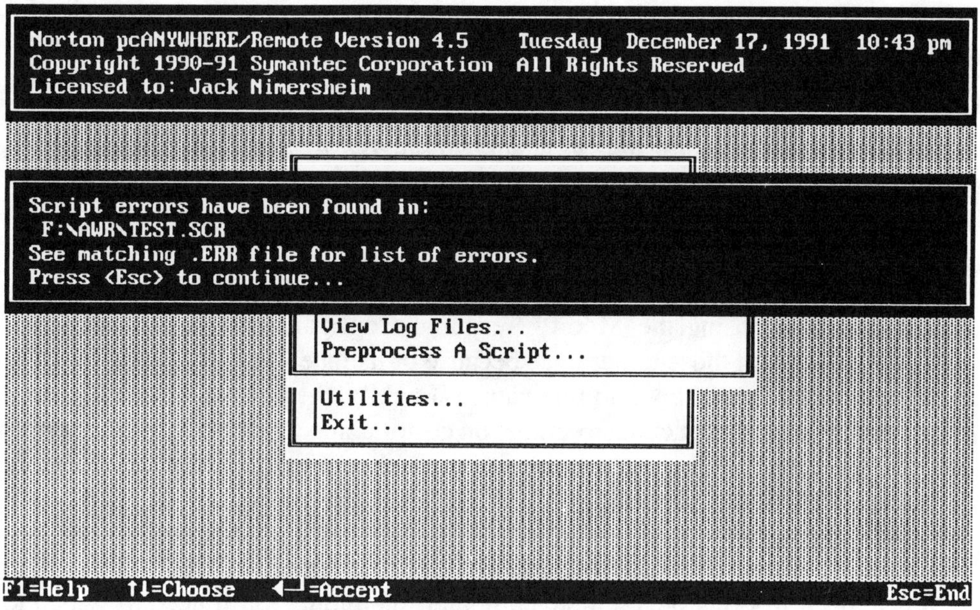

13-17 pcANYWHERE identifies any errors it detects during a script conversion.

Once your SCR script is converted to an SCX file, it can be run using pcANYWHERE/Remote's Run Script option.

Customizing pcANYWHERE's terminal configurations

The last pcANYWHERE utility you're going to look at is also the program's most advanced feature. It actually allows you to modify how pcANYWHERE/Host interprets signals coming from a remote terminal.

When would you want to do this, and why? The answer to the first part of this question is fairly straightforward—you should change a terminal configuration only if you know what you're doing. Getting down to this level of program customization is a highly technical undertaking, and not be attempted by the untrained or faint of heart. As to why you might need to modify a terminal configuration—that also gets a little technical.

Although you've dealt primarily with PC-to-PC connections in this book, almost any kind of computer can be used to access a pcANYWHERE host system. In fact, all you really need to begin remote computing with pcANYWHERE is a "bare bones" terminal and a communications program that allows it to contact the host. Ideally, this terminal should be one of the almost 30 models for which compatibility is built into pcANYWHERE/Host. If this isn't the case, a second alternative is to use a communications program that's able to emulate one of the terminals pcANYWHERE supports.

Should neither a compatible terminal nor terminal emulation be available, don't give up. The situation still isn't hopeless; it just becomes a little more complicated. As a last resort, you can use the Modify Terminal Configurations option on pcANYWHERE/Host's Utility menu to change a preexisting terminal configuration in such a way that it will be able to handle an otherwise unsupported terminal.

Keep in mind my earlier caveat, however. Modifying a terminal configuration is a rather complicated procedure. It requires a considerable level of technical knowledge and no small amount of patience.

Having reiterated this warning, I'm going to outline the general steps used to access and begin using the Modify Terminal Configurations option. Unfortunately, this is one of those topics that, because of its inherent complexity, can't be covered completely in a book of this nature. Luckily, the pcANYWHERE User's Guide contains very explicit instructions on customizing a pcANYWHERE terminal configuration. It's the natural place to look, therefore, should you need more information about this particular pcANYWHERE/Host utility.

Note: Because terminal configurations are used exclusively on the host system, Modify Terminal Configurations isn't available in pcANYWHERE/Remote. If you've been working on a remote computer, therefore, you'll need to switch to your host system for the following exercises.

Modify Terminal Configurations is one of the pcANYWHERE/Host Utility menu options. To access this menu and select the Modify Terminal Configurations option, highlight Utilities on the pcANYWHERE/Host Main Menu and press Enter. Then highlight Modify Terminal Configurations and press Enter.

pcANYWHERE/Host will display a list of the various terminals it supports, as shown in Fig. 13-18. You use this list to choose the terminal configuration you want to modify. The most logical strategy is to pick a terminal that functions similarly to the one you'll be using to contact your host system. Doing so will minimize the number of changes required to make a nonsupported terminal compatible with pcANYWHERE/Host.

For now, highlight Dig Microsystems and press Enter. pcANYWHERE/Host will display the option box shown in Fig. 13-19, identifying the three categories of parameters you can change using the Modify Terminal Configurations utility. These include:

Keyboard Sequences Use this option to modify how pcANYWHERE/Host interprets the signals it receives when selected keys are pressed on the keyboard of the remote terminal.

Screen Sequences Use this option to change how pcANYWHERE/Host controls the display of the remote system. Items you can modify under Screen Sequences include cursor positioning, inserting and deleting lines and characters,

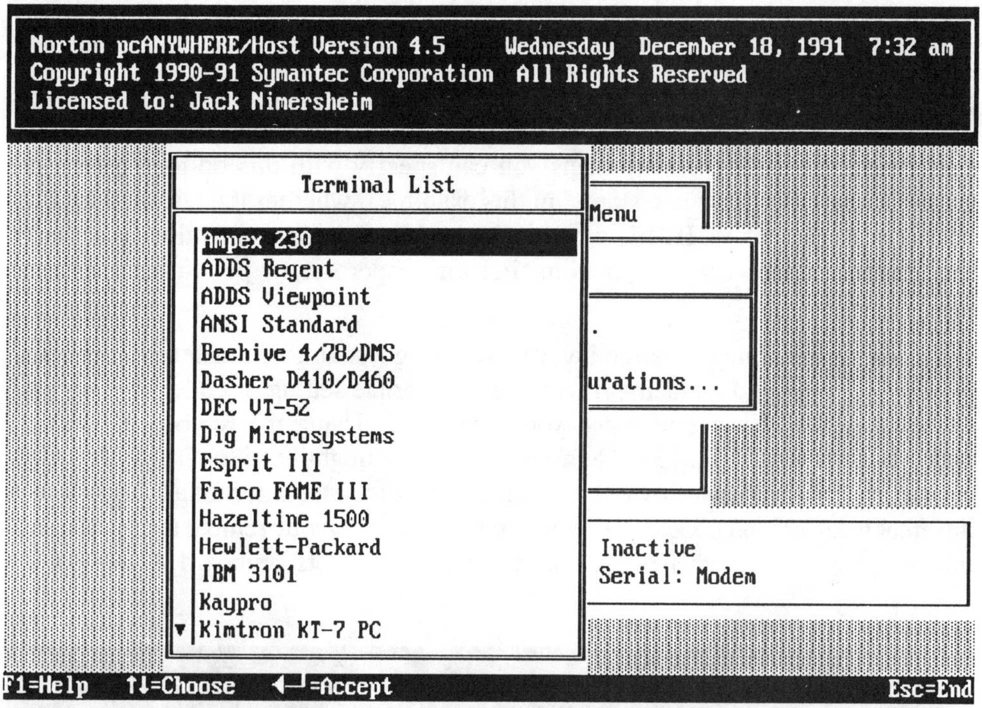

13-18 You can modify the settings for individual terminals supported by pcANYWHERE/Host.

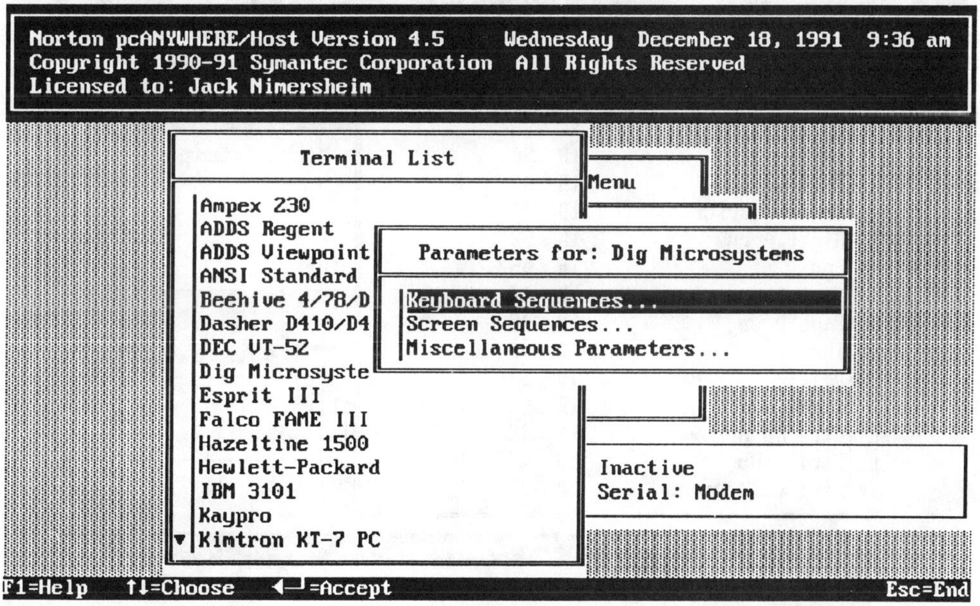

13-19 pcANYWHERE/Host lets you modify parameters in three general categories.

repeat sequences, and display attributes such as highlighting, blinking, underlining, and the like.

Miscellaneous Parameters Use this option to modify those terminal attributes not included in the keyboard and screen parameters. (Hey, it says *miscellaneous*; that's exactly what it means.) Items you can specify with this option include the terminal name (Dig Microsystems, in this example), whether the terminal is compatible with pcANYWHERE, whether its keyboard is case sensitive, whether it requires the row or column position first for proper cursor positioning, and the like.

Typically, terminal modification involves selecting a setting or attribute in the original configuration and replacing it with the appropriate setting or code, often in hexadecimal format, for the terminal you'll be using. Using the Keyboard Function dialog box shown in Fig. 13-20 as an example, you might need to change the original setting for the down arrow key on a Dig Microsystems terminal, 1B 44, to a different hexadecimal code corresponding to the signals your remote terminal transmits to a host system whenever the down arrow key on its keyboard is pressed.

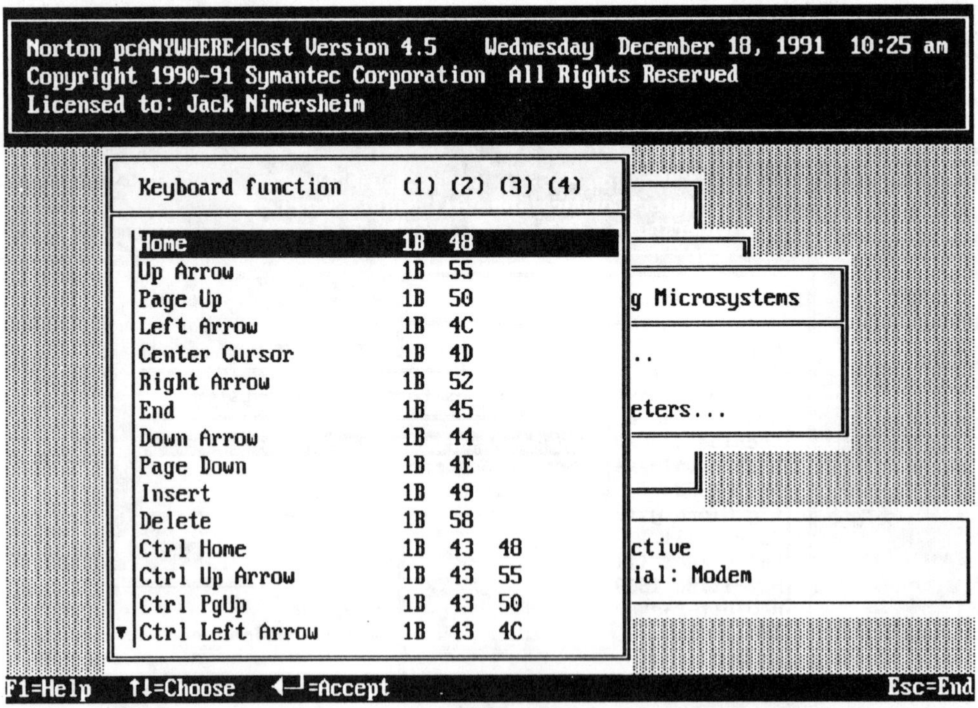

13-20 Modifying a terminal configuration involves changing the hexadecimal codes used to control a nonstandard terminal.

As I said earlier, modifying a terminal configuration requires no small amount of technical knowledge. If you're not sure what you're doing, find someone knowledgeable to help you. The pcANYWHERE documentation provides all the information that either you or someone helping you will need to make virtually any terminal compatible with pcANYWHERE/Host. To exit the pcANYWHERE/Host Utilities and return to the DOS system prompt:

1. Press Esc until you're back at the pcANYWHERE/Host Main Menu.
2. Highlight Exit and press Enter.
3. At the Exit AWHOST prompt, highlight Yes and press Enter.

This closes down pcANYWHERE/Host and returns you to the DOS system prompt. Note: If you've been working on the remote computer, the Exit option on the pcANYWHERE/Remote Main Menu will accomplish the same thing.

Summary

This just about concludes your hands-on examination of pcANYWHERE. All that remains are some troubleshooting tips and techniques, which I'll cover in the next and final chapter.

14
CHAPTER

Tips, techniques, and troubleshooting

Over the past thirteen chapters, I've discussed virtually every major feature of pcANYWHERE. You've learned how to install and set up this extremely impressive PC program. You've looked at how pcANYWHERE's remote-computing capabilities allow you to be, figuratively speaking, in two places at once. You've examined how pcANYWHERE can be used for more traditional on-line activities such as electronic messaging, real-time conferencing, and transferring files between two different computers. You've even seen how to preprogram pcANYWHERE to initiate predefined operations at preset times—without you having to be there to get things started. In short, I've covered, by my estimate, about 98% of pcANYWHERE's capabilities. But I'm still not finished.

What about that remaining 2%—the pcANYWHERE features that didn't fit comfortably within the logical structure I tried to establish for this book? Where am I supposed to cover them? And what should I do with a number of techniques I've stumbled across that, used properly, can improve the overall performance or efficiency of pcANYWHERE? Where do I include information on solving some potential problems you might encounter when you first begin using pcANYWHERE on your PC?

The answer to these questions is simple—to end this book with a chapter like the one you're reading now. I can justifiably discuss an awful lot of miscellaneous information under the title *Tips, Techniques, and Troubleshooting*, which is exactly what I plan to do in this final chapter.

Making the most of memory

There's no denying that the most valuable resource in any personal computer is memory. Almost always in short supply, memory becomes especially critical when you're using a program like pcANYWHERE.

Think about it. Using pcANYWHERE for remote computing actually involves running two programs. First, there's pcANYWHERE itself, in the form of either the pcANYWHERE/Host or pcANYWHERE/Remote module. Like any program, pcANYWHERE requires memory—anywhere from a frugal 40K to a whopping 300K of memory, depending on how you configure and use it. But pcANYWHERE, by itself, is only half the story. Unless you're using pcANYWHERE exclusively to execute DOS commands, an unlikely scenario, you'll also be running application software on the host system—another program, more memory. This same premise holds true on the remote system should you ever need to access the pcANYWHERE/Remote DOS shell.

Clearly, then, any steps you take to increase the amount of memory available on either your host or remote computer will ultimately enhance your pcANYWHERE environment. Allow me to outline a few tricks you can use to either reduce the RAM requirements of pcANYWHERE or, alternately, free additional system memory for pcANYWHERE to use.

Loading DOS in high memory

Beginning with MS-DOS 5.0, DOS itself provides what might be the easiest and most effective method of freeing additional memory for your application programs, including pcANYWHERE. Whereas previous versions of the PC world's most popular operating environment consumed anywhere from 40K to 60K of precious conventional RAM—that is, RAM below the so-called 640K DOS barrier—MS-DOS 5.0 can load a portion of itself above the 640K memory address.

On an 80286 system with at least 64K of extended memory, adding the following two lines to the beginning of your CONFIG.SYS file will increase by approximately 45K the amount of conventional RAM available to your application programs:

```
DEVICE=HIMEM.SYS
DOS=HIGH
```

As an added attraction, loading DOS into high memory will automatically cause it to place any buffers you create in CONFIG.SYS into high memory, also. Each DOS buffer you create consumes just under 500 bytes of memory. If you normally use 30 buffers on your system, therefore, the DOS=HIGH command will release approximately 16K of conventional RAM for other uses.

The news is even better for those of you who own a 386 or better system with extended memory. You should begin your CONFIG.SYS file with the following three lines:

```
DEVICE=HIMEM.SYS
DOS=HIGH
DEVICE=EMM386.EXE RAM
```

Line 3 in this example loads a special DOS memory manager, EMM396.EXE, which is designed to convert any extended memory your system has to expanded memory. Furthermore, adding the RAM parameter to this line causes EMM386.EXE to configure high memory to accept any device drivers or terminate-and-stay-resident (TSR) utilities you might use. If you recognize that term, TSR, give yourself a gold star. I used it in an earlier chapter to describe how the resident portion of pcANYWHERE/Host works.

The next logical question then becomes, "Can someone with a 386 or better system load a portion of pcANYWHERE/Host into high memory?" The answer is, "Yes, as you'll see in the next section."

Loading pcANYWHERE/Host in high memory

Entering the appropriate command in the Command to Load Host into High Memory field of the General Preferences dialog box tells pcANYWHERE to load portions of pcANYWHERE/Host into high memory. On a 386 or better system with DOS 5.0 set up as outlined in the previous section, this command is:

```
COMMAND /C LH
```

A second popular 386 memory manager that can be instructed to load TSRs into high memory is Quarterdeck's QEMM386. If you're a QEMM user, your Command to Load Host into High Memory field should contain the following command:

```
LOADHI
```

To be totally effective, both of these techniques require that you have a 64K block of RAM available in high memory. Even if your available memory is less than this, however, loading a portion of pcANYWHERE/Host into high memory will still increase the amount of conventional RAM available to your application programs. My advice, therefore, is to use the Command to Load Host into High Memory field if you own a 386 or 486 PC. It can only help.

Disabling the Maximum Display Speed option

By default, pcANYWHERE/Host compresses screen updates it transmits to the remote system when running DOS applications. While data compression improves the performance of programs running on the host, it also increases by almost 50K pcANYWHERE/Host's memory requirements. Disabling data compression frees this memory up for other uses. To disable data compression:

1. Select Preferences from the pcANYWHERE/Host Main Menu.

2. Select Features from the Host Preferences menu.
3. Highlight Maximize Display Speed in the Host Features Preferences dialog box and press Enter.
4. Highlight No in the subsequent option box and press Enter.
5. Press Esc to close the Host Features Preferences dialog box.
6. Choose Yes at the Save Changes prompt.
7. Press Esc to return to the pcANYWHERE/Host Main Menu.

This simple operation will free approximately 46K of memory on your host system. Note: Although helpful in almost any situation, disabling data compression might be a problem if you see the following message when trying to set up pcANYWHERE/Host to accept incoming calls from a remote computer:

Unable to load high speed graphics overlay—not enough memory.

If this happens, and if you've already taken all other possible steps to provide the largest amount of memory you can to pcANYWHERE/Host as outlined previously, you'll need to disable data compression in order to set up your pcANYWHERE host system.

Modem-related problems

Whereas, in many cases, a lack of memory represents a mere inconvenience, modem-related problems can bring pcANYWHERE to a screeching halt. This shouldn't surprise you. A modem is, after all, the single, most critical piece of equipment on any PC set up for telecommunications. Complications can arise in any one of several places along your pcANYWHERE connection, including:

- cables and other connections
- the serial port to which a modem is connected
- the modem itself
- the telephone line

Let's discuss some common problems associated with each of these items.

Making the right connections

Allow me to define humiliation for you—humiliation is when you spend several hours seeking a high-tech solution to communication conflicts, only to discover that the problem was a loose cable or other connection. This happens to everyone at least once, so there's no cause to feel overly embarrassed when it happens to you. The first thing you should check, therefore, should pcANYWHERE have trouble either initializing your modem or placing calls, are the critical links that connect your PC to the outside world.

The specific connections you need to check are the RJ11 telephone cord that connects the modem to the wall plug and the serial cable that runs between your

PC's serial port and an external modem. The serial cable is most suspect if pcANYWHERE isn't able to initialize your modem when you attempt to establish a remote connection. The RJ11 cord would be the prime culprit if, after initialization occurs, problems arise either accepting incoming calls with pcANYWHERE/Host or placing outgoing calls with pcANYWHERE/Remote. Trust me, you'll save yourself a lot of trouble (as well as potential humiliation) if you check these connections at the first sign of trouble.

Resolving serial port conflicts

As I've stated several times throughout this book, a serial port (or COM port) resembles an electronic doorway through which signals pass to and from your PC and modem. This doorway metaphor also provides insight into how serial port conflicts can cause problems for pcANYWHERE.

Assigning the proper COM setting

Like most homes, your PC probably has more than one doorway leading to the outside world. Today's systems commonly have at least two serial ports (COM1 and COM2). Some include three or four (COM1 through COM3 or COM4). For pcANYWHERE to work properly, you need to make certain that it's indeed using the serial port to which your modem is attached.

Assigning an incorrect COM setting to pcANYWHERE is a lot like trying to unlock your front door with the back-door key. It won't work. If pcANYWHERE and your modem have trouble communicating with one another, therefore, the first thing to check is whether they're set up to use the same serial port, or COM number. Both pcANYWHERE/Host and pcANYWHERE/Remote store COM information in their individual hardware configurations.

Double trouble

Problems will also arise if you assign multiple devices to the same COM port. While it's virtually impossible to connect two cables to the same physical port, this could happen, for example, if you install an internal modem configured to use COM1 to a system that already has an external serial port assigned the same COM number. Imagine two people trying to get through a narrow doorway at the same time and you'll have some idea of the potential hassles this can cause.

It's impolite to interrupt

Problems can arise between two devices even when they're assigned to separate COM ports. Should this happen, the source of the problem is likely to be an interrupt, or IRQ, conflict.

pcANYWHERE needs to know two values in order to communicate successfully with your modem. The first, which I've already discussed, is the COM port to which it's attached. Just as critical, however, is the interrupt address your PC uses internally to coordinate communications with this COM port.

By default, DOS uses only two interrupts, IRQ3 and IRQ4, to control its first four COM ports, COM1 through COM4. To confuse matters further (as if you really needed this), IRQ4 controls COM1 and COM3, while COM2 and COM4 are handled by IRQ3.

As a rule, an interrupt conflict won't rear its ugly head unless and until two devices actually try to use the same interrupt at the same time. (This differs from serial port conflicts, where just assigning two devices to the same COM port can be disastrous.) Consequently, it's possible to use, for example, both COM1 and COM3 on your system without changing their interrupt addresses, so long as the two devices involved are never running concurrently. A serial printer using COM1, for instance, can coexist quite peacefully with a modem connected to COM3, providing you don't try to use this printer during an on-line session. If you do, chances are your system will crash and require a cold boot to recover.

That's the bad news. The good news is that pcANYWHERE allows you to change the IRQ setting associated with a given COM port, thus eliminating potential conflicts before they occur. To modify an interrupt setting:

1. Select Hardware Configurations from the pcANYWHERE/Host or pcANYWHERE/Remote Main Menu.
2. Highlight the configuration you want to modify and press Enter.
3. Highlight the Device/Port field in the Configuration dialog box and press Enter.
4. Highlight Serial in the subsequent option box and press Enter.
5. Highlight Custom on the subsequent list of serial ports and press Enter.

This will display the dialog box shown in Fig. 14-1, which you can use to specify the IRQ and Address settings. The documentation for your internal modem or serial card should contain information on which interrupts it supports. Specifying a different interrupt generally involves changing a dip switch or jumper pin on the board itself, then modifying the IRQ and Address fields in pcANYWHERE to match your new settings after making these modifications:

1. Press Esc three times to exit the Hardware Configurations option.
2. Choose Yes at the Save Changes prompt.
3. Press Esc to return to the Main Menu.

The above steps, if performed properly, will resolve potential conflicts that can arise from two devices trying to share the same interrupt settings.

Modem madness

If cables or communication settings aren't the source of your problems, the next logical place to check is the modem. Once again, where the difficulty lies can

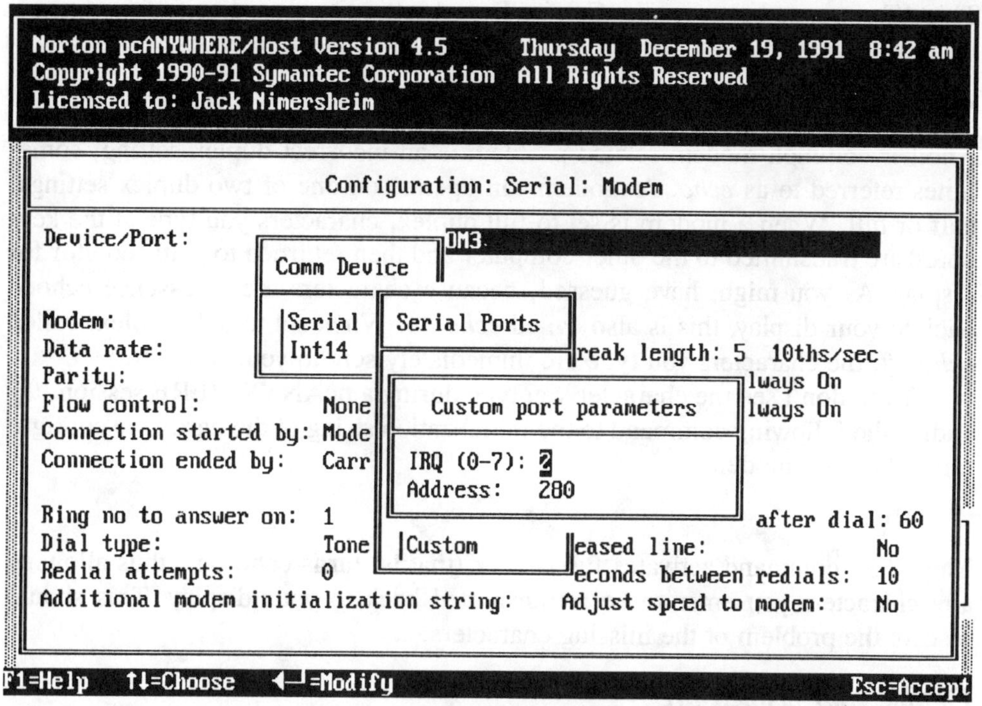

14-1 Specifying custom port parameters lets you eliminate potential IRQ conflicts in pcANYWHERE.

often be determined by how a problem manifests itself. In other words, understanding the symptoms will help you diagnose the disease, if you will.

The constant connection

The key to establishing successful communications between pcANYWHERE and your modem is to coordinate these two critical elements of your on-line environment. One indication that a conflict exists between pcANYWHERE and your modem is if pcANYWHERE/Host falsely reports that it's connected to a remote system when no such connection exists. The most common cause of this problem is the presence of a false Carrier Detect (CD) signal.

Many modems ship with Carrier Detect turned on as their default setting. CD should be turned off for most pcANYWHERE operations. If you experience false connections, check to see if a dip-switch setting on your modem allows you to turn Carrier Detect off, so that a CD signal exists only after an actual connection between the host and remote systems has been established. If you can't manually change the CD setting, an alternate solution to this problem is to include the following command in the initialization string of the hardware configuration for this modem:

```
AT&C1
```

This AT command changes the Carrier Detect setting on a modem to false, meaning it will show a CD signal only when an actual connection exists.

Silent partner

Another common modem-related problem is an incorrect duplex setting, sometimes referred to as *echo*. A modem can operate on one of two duplex settings, half or full. When a modem is set to full duplex, characters you type at the keyboard are transmitted to the other computer and then returned to your monitor for display. As you might have guessed, because characters are in essence echoed back to your display, this is also called *echo on*. When set to half duplex, called *echo off*, the characters you type are immediately sent to your monitor.

If you don't see the character you type during a pcANYWHERE session, try adding the following command to the initialization string of the hardware configuration for your modem:

```
ATE1
```

The ATE1 command activates full duplex (that is, turns echo on), thus allowing any characters you enter to be retransmitted back to your display. This should resolve the problem of the missing characters.

Double your displeasure

Of course, the opposite condition of the problem outlined above is when everything you type appears twice on your monitor, making the display llooookk lliikkee tthhiiss. The best way to fix this is to turn echo off by adding the following command to the initialization string of the hardware configuration for your modem:

```
ATEO
```

The ATEO command deactivates full duplex—or, in essence, turns echo off—thus preventing the other computer from returning the second occurrence of each character to your monitor.

Hold that call

Everyone loves their telephone. And more and more, everyone loves to enhance their phone service by adding all kinds of features to it. There's call forwarding, conference calling, and, perhaps the most convenient of all phone enhancements, call waiting. I don't know about you, but busy signals drive me crazy. Like most people into instant gratification, I like nothing better than reaching people immediately once I decide to "reach out and touch" them. From this perspective, I truly appreciate call waiting. The same convenience that I cherish so much for voice communications, however, caused me no small amount of grief whenever I tried to telecommunicate—until I found a way to stop singing *The Call-Waiting Blues*.

The call-waiting blues

That innocuous little beep that heralds an incoming call on lines equipped with call waiting can wreak havoc on an unprepared modem connection, often knocking you totally off-line in less time than it takes to bemoan your bad luck.

The good news is that many telephone companies now allow you to temporarily disable call waiting by preceding the number you're dialing with a special code, generally either *70 or 70#. With rotary phones, which lack the asterisk (*) and pound sign (#) keys—or, alternately, on older, pulse-dialing telephone systems—you might be able to accomplish the same thing with the numeric sequence 1170.

To see if call waiting can be disabled on your phone system, simply pick up your phone and try entering the codes listed above. If you hear a second dial tone after any of them, then your local system supports this feature. By adding the appropriate code in the Prefix field of your Host or Caller Information files, you can prevent call waiting from prematurely terminating a pcANYWHERE session.

Call-waiting blues, the refrain

If none of the codes listed above solve the call waiting conundrum on your phone system, don't give up hope. A second way to avoid this potential problem is to use the AT SO command to increase your modem's lost carrier to hang-up command setting to a higher number. To defeat call waiting using this technique, add the following command to the initialization string of the hardware configuration for your modem:

```
AT S10=255
```

This command will increase the amount of time your modem will tolerate a lost signal before dropping a connection to 2.5 seconds. That's more than enough time to ensure that the brief call waiting signal will no longer knock you off-line. True, you'll probably have to put up with some garbled data, as call waiting announces an incoming call, but at least your pcANYWHERE connection will survive.

Money matters

What better way to end this book than a practical discussion about a pragmatic topic? Specifically, let's talk about saving money. As I've mentioned in passing several times throughout this book, telecommunicating is a "pay as you play" proposition. Every time you sign on to a commercial information service such as CompuServe or Delphi, someone, somewhere, is ticking away the minutes you spend on-line and calculating how much you'll have to cough up based on a per-minute access charge. Piling up long-distance costs or, alternately, an interstate tariff on top of the standard connect charges levied by commercial on-line services only exacerbates the situation. Forget this fact for even a short time and you'll quickly find yourself faced with a monthly phone bill exceeding the GNP of many third-world countries. While it's impossible to completely eliminate the cost of telecommunicating, there are steps you can take to reduce it as much as possible.

Aesthetic, yes; economical, maybe

With your indulgence, allow me to begin this discussion with a personal story. A little over two years ago, the Nimersheim clan pulled up roots and relocated. Abandoning the hectic pace of urban life, my wife, my son, and I became the newest residents of Vevay, Indiana—an idyllic little burg nestled against the banks of the beautiful Ohio River, about 50 miles northeast of Louisville, Kentucky. I don't mention this fact merely to boast about our new surroundings; shifting locales as we did also prompted a major rethinking of how I managed my on-line activities. Let me explain.

As recently as ten years ago, the whole idea of working anywhere other than a major metropolitan area would have seemed ludicrous. Today, however, one city or town with access to a post office, delivery service, and telephone line is pretty much like any other place having access to the same resources. Admittedly, FedEx might have to travel a little farther to absolutely, positively deliver something to my rural community overnight, but they still do it—and do it reliably. That takes care of transporting the physical materials I (or anyone else, for that matter) need to work. But what method is available to deliver the intellectual fruits of my labors—the articles, reviews, columns, manuals, books, and so forth that I write about those physical components I receive? These, after all, often need to be turned around even more quickly. For that, I depend on telecommunications, the very subject I've discussed in such great detail over the past thirteen chapters. Here, once again, as I've pointed out several times, proximity to a major city accounts for nothing. Well, almost nothing, a point that leads directly to that promised discussion about the potential cost of telecommunicating.

Foreign affairs

Alas, it pains me to report, not even our idyllic burg was perfect. For all its aesthetic advantages, living in a small town like Vevay does have one major drawback—it's a long-distance call from everywhere, including any city possessing local access numbers to the various commercial services I use regularly for my electronic correspondence. Consequently, after relocating to Vevay, I faced the unenviable task of paying the extra costs associated with piling long-distance charges on top of the normal connect charges for using a commercial information service. One potentially promising discovery I made during the transition to rural life was that, depending on which rural area you inhabit, there might be a way to minimize these long-distance costs. The secret is installing something called a *foreign exchange*.

Basically, this entails having your phone company request a special line and corresponding exchange from its counterpart in a larger, nearby city, which it then assigns as your local phone number. The end result is that people dial your phone just as if they were placing a call to the larger city. For example, rather than using the 812 southern Indiana area code to reach me in Vevay, callers dial a number in the 513 area code assigned to Cincinnati, the metropolitan area nearest to my location. Similarly, I can make local calls to Cincinnati numbers, just as if I lived there. (The down side to all this is that, conversely, dialing up a next-door neighbor in my new home is billed as a long-distance call due to my possessing a foreign exchange. But that's another discussion entirely.)

Of course, nothing in life is free. You must pay an additional monthly fee to obtain a foreign exchange, the amount of which varies from phone system to phone system. If this charge is less than the long-distance fees you would normally run up each month for your online activities, however, the net result will be a reduced phone bill. As an added attraction, because the foreign-exchange surcharge is a fixed fee, it's possible to more easily project your monthly telephone expenses. You should be aware, however, that setting up a foreign exchange is not always this straightforward. Again, depending on where you live, there might be hidden costs.

Remember when AT&T provided long-distance service for the entire country? No longer. Congress breaking up Ma Bell into several "baby bells" changed all the rules for long distance. Under the post-AT&T setup our beloved bureaucrats created, each long-distance phone carrier is assigned its own local-access and transport area (LATA), a specific geographic area for which it, and it alone, is responsible. In my part of the country, for instance, Cincinnati Bell controls a LATA that includes parts of Southwestern Ohio, Southeastern Indiana, and Northern Kentucky. A foreign exchange can't cross over LATA lines, so mine would have to be a Cincinnati number. I couldn't request one for Louisville or Indianapolis, two other major cities in close proximity. Normally, this wouldn't present a problem because Cincinnati provides local-access numbers to all the online services I use. But have I ever expected my life to be normal? Of course not. If there's a problem to be encountered, I'll encounter it, as I did when I inquired about getting a Cincinnati foreign exchange for our new home.

Keep in mind that AT&T was a national communications network. As such, it had been granted some pretty exclusive privileges down through the years, privileges that in turn allowed it to operate efficiently on a national scale. The current situation is quite different. Being smaller, each of the "baby bells" is governed primarily by the trade and commerce laws of the state in which its corporate offices reside. Somewhere along the way, some faceless bureaucrat decided that this created a situation in which an interstate tariff could be levied each time a local phone company's trade or commerce crossed state lines. And there's the rub, as Shakespeare once wrote, to procuring a foreign exchange number in certain locations— my own home town of Vevay, to name but one.

Unfortunately, there's a state border between Vevay and Cincinnati. Consequently, when I inquired about setting up a foreign exchange for my new home, a polite Cincinnati Bell representative informed me that I would be charged our faceless bureaucrat's interstate tariff, to the tune of 10 cents per minute, on each call placed through this number.

Clearly, a foreign exchange isn't the way to go for me. If you're in a rural area governed by the same LATA as a major city in the same state, however, it might be one method you can use to control your long-distance phone bills. At the least, it's worth checking out. If the foreign-exchange route isn't right for you, don't despair completely. There are other ways to keep expenses at a minimum while enjoying the maximum benefits of pcANYWHERE.

Timely calling

One good money-saving strategy is to place any calls you make to a pcANYWHERE host, commercial service, or BBS at a time of the day when long-distance rates are lowest. Unless you're participating in a scheduled real-time conference, the hour at which you call another computer is totally irrelevant. 12:00 noon and 12:00 midnight are virtually synonymous to one another as far as two computers are concerned. Consequently, you can save a bundle by scheduling most of your on-line activities for "discount hours"—between 11:00 P.M. and 8:00 A.M. during the week or on weekends—thus taking advantage of long-distance rates that average 65% lower than those charged during normal business hours.

State of confusion

While we're on the subject of long-distance charges, let me reveal another little-known fact that might help further reduce your on-line expenses. One basic irony of our current convoluted nationwide telephone network is that interstate rates are generally lower than those levied on long-distance calls placed within the same state. Consequently, if you need to contact commercial services long distance, placing a three-minute call to CompuServe, for example, through the local SprintNet number for a city near your own but across a state border is usually less expensive than if you make that same three-minute connection through a network node in your home state. Don't ask me to explain the logic behind this illogical rate structure because there doesn't seem to be any. Just accept it and, like any good penny-pincher, take advantage of it if possible. Over several months, if you're forced to use long distance, the savings generated by calling your favorite on-line service over an interstate rather than an intrastate connection can be considerable.

Tag-team calling

I'm almost but not done examining ways to save money making long-distance calls. To close this discussion out, I'd like to share an electronic missive I received shortly after sharing the long distance woes I described above with the readers of one of my monthly magazine columns. This particular message was waiting for me one Saturday morning when I signed on to BIX. (Since my move to Vevay, I now make it a habit to call when lower, weekend long-distance rates are in effect.) Truth to tell, I had to read it several times before I truly comprehended all the ramifications of what its sender had accomplished.

Stated simply, this reader "beat the system," taking full advantage of the quirks associated with how our telephone system is structured to substantially lower his monthly phone bill. And so, again with your kind indulgence, I'm going to reprint the salient portions of that electronic message. I love what this guy did! I think you will, too.

> Hi Jack,
>
> I just read your Connections column in the July issue of *Computer Monthly* about the problems of accessing on-line services from remote areas. Facing a similar situation, I thought I'd share with you a creative way I found of getting around this particular problem.
>
> Regarding toll calls: I wanted to tell you about a couple of things that I have done to get around them.
>
> I am a heavy user of CompuServe. However, the closest CIS (Information Service) node to me was a "short-distance" toll call, on the order of 7 cents per minute or thereabouts. That added up fast! To try to reduce this, I asked my local phone company, New Jersey Bell, about what they call *selective calling*, a relatively new service available in selected areas.
>
> It turned out that, for about $3 or $4 a month, I could get 20 hours of calls to Princeton (which is where the CIS node is) from my home. Additional minutes after 20 hours are billed at slightly under 5 cents per minute. This deal sounded a lot better than standard long distance, so I signed up. I was blissfully happy until I got my first month's bill and discovered that I had squandered my base 20 hours and spent about another six thousand minutes taking advantage of what I thought was going to be a real money saver. I didn't blame New Jersey Bell for my financial dilemma, mind you. I had merely grown complacent and not bothered to keep track of my on-line charges.
>
> After bemoaning my situation to another CompuServe user, I was

given a wonderful idea that solved all my long-distance woes. I realize it won't work for everyone, but in some cases it can save people a lot of money on long-distance calls.

What I did was check my phone book for all the local dialing areas that can be accessed from my phone exchange. In doing so, I discovered a town about halfway between my home and Princeton that's a local call for me. Furthermore, from this town it's a local call to Princeton.

Now I happened to have a friend who lives in an apartment in this "halfway" town. So, with my friend's permission, and a key to the utility room in his apartment complex, I had the phone company install a line to his address that would be billed to my name. I also asked that this line have call-forwarding activated on it.

For a minimum installation fee of $42, the phone company was more than willing to run a line to the phone panel in the utility room and put in an interface box with a single, modular test jack on it. After this jack was installed, I simply plugged a cheap phone into it, dialed in the proper code to forward all calls received at this new number to the Princeton CompuServe access number (a local call, remember), and then unplugged the phone. Finally, I detached the wires inside the modular test jack, just to guarantee no one could make calls from this new number with their own phone, and went home.

The end result? I can now make unlimited calls to CompuServe for a total of $14.44 per month—the cost of the basic service for the line, the call forwarding, and New Jersey's new sales tax on telecommunications services. Compared to the $250 to $350 a month I was paying in monthly phone bills for my on-line activities prior to then, it was worth the $42 installation fee, the hassle of bugging my friend, and the inconvenience of having to wait out in the freezing cold for the installation man to show up.

Like I said, this won't work for everyone, but if it can save a few people the kind of bucks it saved me, it might be worthwhile to pass along.

I have to tell you, the solution this reader finally discovered to overcome his long-distance woes is so creative and so obvious that I'm kicking myself for not thinking of it first. (Not that it would have helped me personally, because there are no "halfway" towns between Vevay and Cincinnati like the one my canny correspondent discovered between his house and Princeton.) And, best of all, everything he did was legal. In fact, he told several phone company representatives exactly what he was planning to do and why, and nobody he talked to seemed to care that they stood to lose out on a few hundred of his dollars each month.

To figure out how to beat the phone company at its own game! And to do so legally, to boot! Ladies and gents, what we have here is true genius at work. Whoever originally stumbled upon this little bit of long-distance sleight of hand deserves a medal.

Parting words

This joyous note (pun intended) strikes me as the ideal place to wrap things up. Fourteen chapters have come and gone and you've finally completed your grand tour of pcANYWHERE. You've learned what pcANYWHERE is, what it does, and how it works. I hope you've enjoyed your little journey, and I suspect you can't wait to explore the amazing world pcANYWHERE places within your electronic reach. Go for it.

One final, personal note: If we should ever run into one another on-line, be sure to say Hi! Making new friends is one of the nicest things pcANYWHERE makes possible.

A
APPENDIX

What's out there?

Being able to use pcANYWHERE to access another computer doesn't do you much good unless you know what's out there and, even more crucial, how to get in touch with any commercial services and local BBSs you decide to contact. This appendix is designed to provide such information for some of the literally thousands of commercial and public systems that are waiting for you to call.

What's in this appendix?

The specific information included for a given system depends to a large part on the type of system it is. In the case of local BBSs, you'll find their access number (including area code), along with the line settings you should use when calling them and, whenever appropriate, a brief description of the type of services they offer. For commercial on-line services, in addition to the items listed above, I've also provided a number you can call for more information on a given service. As a rule, this number will put you in contact with another living, breathing, warm-blooded human being who is prepared to answer any questions you might have about the commercial service he represents. Given that these are "commercial" ventures, this person will also be a sales representative and know how to sign you up for that service, should you decide to do so.

These listings represent some of the resources you'll need to get up and running when using pcANYWHERE for anything other than a pcANYWHERE/Host-to-pcANYWHERE/Remote connection. Enjoy.

Commercial on-line services

First, a caveat—commercial on-line services come and go. Some have been around for years and are still going strong, others are new and untested, while others still have appeared for a short while and then faded from sight, victims of the same market influences that govern any commercial endeavor. The on-line services listed here were seemingly healthy, profitable and, therefore, still active as of December, 1991, when I completed this book. If some have since gone "belly-up" or new ones surfaced, then my list will be unavoidably out of date or incomplete. All I can do is extend an apology, before the fact, for any confusion and/or inconvenience these inaccuracies might cause, and offer the explanation that such changes are inevitable in the dynamic world of telecommunications.

CompuServe
5000 Arlington Centre Blvd.
Columbus, Ohio 43220
800-848-8199 (In Ohio and Canada, 614-457-0802)

Arguably the most successful commercial services currently in business. In 1989, CompuServe "absorbed" The Source—one of the first on-line services to go into business and, until the recent merger, CompuServe's chief competitor. Consequently, CompuServe now claims over 900,000 subscribers world-wide.

A partial listing of services CompuServe offers includes: electronic messaging and a variety of e-mail "gateways" through its CompuServe Mail service; easy access to several commercial databases for on-line research; an electronic shopping mall; the ability to upload and download public domain software and shareware for numerous types of personal computers and operating systems; real-time conferencing; and literally dozens of user forums dedicated to such areas of special interest as comic books, gardening, science fiction, medicine, technology, and more. Reflecting a recently developing trend, several manufacturers of PC hardware now offer on-line technical support through special CompuServe forums. CompuServe also allows you to send and receive text-only (that is, non-graphic) FAX messages using CompuServe Mail.

CompuServe recently revised its pricing structure to provide many of its more popular services for a basic monthly fee, rather than charging per-minute connect costs as it has in the past. For $7.95 per month, subscribers can now access a wide range of on-line services, including electronic mail, Associated Press Online, the Grolier Academic American Encyclopedia, stock quotes, weather information, and a number of games and quizzes. Included in this basic rate is the ability to send a limited number of e-mail messages, approximately 60 per month. Users who send more messages in any given month will be charged for their extra e-mail activity based on the number and size of the additional messages transmitted.

DELPHI
General Videotex Corporation
Three Blackstone Street
Cambridge, MA 02139
800-544-4005 (In Massachusetts, 617-491-3393)

Like CompuServe, DELPHI offers a wide range of options, all available at reasonable rates. Unlike some commercial services, DELPHI doesn't charge an annual membership fee, so you pay only for the actual time you spend on-line. Recently, in addition to lowering its hourly access fee to $6 an hour, DELPHI introduced the 20/20 Advantage Plan, a special billing arrangement that allows you to spend your first 20 hours each month on-line for the minimal fee of $20 per month. Electronic mail, special interest groups (SIGs), support forums, research databases, and real-time conferences are only some of the services available on DELPHI. (If you do decide to subscribe to DELPHI, drop by the Writer's Conference I've mentioned previously in this book some Thursday night around 9:30 p.m., EST.

For on-line access to DELPHI and subscription information about this service:

- Set your modem to 300/1200/2400-baud, E,7,2.
- Dial 1-800-365-4636.
- Once a connection is established, press Enter once or twice to "wake up" the DELPHI system.
- At the USERNAME prompt, type JOINDELPHI and press Enter.
- At the PASSWORD prompt, type ANYWHERE and press Enter.

The option of whether to actually subscribe on-line to DELPHI is totally your choice. If you do subscribe and for some reason are dissatisfied with DELPHI, the company offers a money-back guarantee.

Dow Jones News/Retrieval
P.O. Box 300
Princeton, NJ 08543
609-542-1511

If business is your game, then Dow Jones is the name of a service you should investigate. Yes, Dow Jones, mainstay of the Wall Street Journal financial section, also sponsors an on-line commercial information service. Needless to say, this particular offering has a decidedly business-oriented slant. For example, Dow Jones gives you fast access to up-to-the-minute stock quotes for both domestic and international exchanges, 24 hours a day. It also includes a number of on-line reports and, in some cases, actual newspapers, ranging from *Historic Dow Jones Averages* to *Standard & Poor's Online* to excerpts from *The Wall Street Journal*, to actual transcripts of the popular television show, *Wall Street Week*. Dow Jones is primarily targeted toward business applications; however, it's not exclusively so. It also includes a shopping service and provides access to airline schedules, information on colleges and universities, movie and book reviews, and a number of other, less corporate, activities.

BYTE Information Exchange (BIX)
One Phoenix Mill Lane
Peterborough, NH 03458
800-227-2983 (In New Hampshire, 603-924-7681)

BYTE is a very well-known and popular computer magazine, having been published in one form or another for over 15 years. BIX is the on-line service sponsored by the same folks responsible for *BYTE*. It provides, quite simply, a wealth of information on computer products and related technologies. You can, for example, submit questions to and get answers from regular contributors to *BYTE*. BIX also sponsors regular conferences on a wide range of computer-related topics. Beyond such high-tech stuff, however, BIX also offers a number of other general services, similar to those available from CompuServe and DELPHI—electronic mail, program libraries, vendor support, research databases, and so forth. I like BIX and sign onto it often, as will anyone who's seriously into computers.

Computer Monthly Electronic Edition (CMEE)
One Chase Corporate Drive, Suite 300
Birmingham, Alabama 35244
205-988-9706

This is the service I used in chapter 10 to introduce you to pcANYWHERE's Quick Connect feature. Like BIX, CMEE is sponsored by a computer magazine, *Computer Monthly*. It, too, will take questions and provide answers from regular contributors to this popular publication—including me, because I write two monthly columns for *Computer Monthly*. CMEE doesn't charge an access fee for the time you spend on-line. People living outside the metropolitan Birmingham area, however, will have to pay for a long-distance phone call when contacting this particular service. You can register directly with CMEE the first time you call that system, using one of the following access numbers:

1-205-655-4059 (2400-baud)
1-205-655-4065 (9600-baud or any slower baud rate)

GEnie
401 N. Washington St.
Rockville, Maryland 20850
800-638-9636

If you think Delphi is a deal, wait until you hear about GEnie. This popular commercial information service recently reduced its monthly fee to a mere $4.95 for unlimited nonprime-time access to over 100 selected products and services. With DELPHI, Prodigy, and GEnie now offering comprehensive access to their on-line offerings for a flat monthly fee, it appears that modem communications are starting to lean toward the same product/pricing structure used for years by cable TV companies.

GEnie offers the standard fare usually associated with a commercial service. Electronic messaging, real-time conferences, file transfer, on-line support for specific hardware and software products, electronic shopping, special-interest round tables; all are available to GEnie subscribers. Most are included in that $4.95 monthly fee. Though not as easy to use as Prodigy—GEnie still relies on text menus and a somewhat obscure command structure to get you from here to there—GEnie's rock-bottom access fee nevertheless makes it an attractive alternative for anyone wanting to test the waters of commercial on-line services. For a little investment you can learn a lot.

Electronic mail services

Electronic mail (or e-mail) operations resemble commercial on-line services in many ways. You sign on to an e-mail service much as you would CompuServe or DELPHI—that is, by entering your assigned user ID and password. They do, however, tend to be more narrowly focused in the types of services they support, concentrating primarily on offering an electronic equivalent of the U.S. Postal Service. Some electronic mail services can be accessed through commercial on-line services, and some will allow you to access such services through special "gateways." It's possible, for example, to send a message to someone's MCI Mail address from CompuServe Mail, and they, in turn, can reply back to your CompuServe user ID.

Businesses tend to like e-mail because it offers the immediacy of an overnight delivery service like Federal Express at a fraction of the cost. One caveat to using e-mail is that, as a rule, they tend to be awkward, at best, if you need to transmit anything other than text files. At worst, certain e-mail services are simply incapable of handling other file types. You should consider this limitation before deciding on which e-mail service will meet your needs.

As is the case with commercial on-line services, the three companies profiled here do not represent a complete list of e-mail services.

AT&T Mail
P.O. Box 3503
New Brunswick, NJ 08903
800-367-7225

AT&T Mail charges an annual fee of $30 and, in addition, you pay for each message sent, with this cost dependent upon message length. Like many e-mail services, AT&T does offer an 800 number for actual connection, thus saving you long-distance charges to compose and deliver mail.

AT&T's MailFax lets you transmit FAX messages to anyone with a standalone FAX machine or, alternately, FAX capabilities on their PC. One big advantage to using MailFax is that you can use it to FAX graphic files, providing you've purchased AT&T's proprietary Access Plus software.

MCI Mail
1150 17th Street NW, 8th Floor
Washington, DC 20036
800-444-6245

And in this corner . . . MCI. Yes, the two biggest telephone companies are also squaring off against one another in the e-mail market. MCI's annual fee is slightly lower than that of AT&T—$25 vs. $30. Its per-character charge to actually send messages, however, is slightly higher, at least for shorter missives. In tandem, these two items pretty much cancel each other out.

MCI's FAX service, Fax Dispatch, is not currently set up to allow the transmission of graphic images. Fax Dispatch does, however, let you register a copy of your letterhead for inclusion in a FAX transmission. MCI Mail's $10 a month volume user's billing option is one of the better bargains available on any dedicated e-mail system. Like AT&T, MCI offers toll-free, 800-number access for its users.

Local BBSs

These are the guys who really started it all. Long before there was CompuServe, years before AT&T ever got into e-mail, individual hackers started hooking up modems to their personal computers and providing an on-line forum for PC users to pursue their shared interests. If not for thousands of dedicated system operators (sysops) throughout the country—indeed, the world—back in the early days of personal computers, there might not even be a software category called telecommunications. Even if there were, in all likelihood it wouldn't be as prevalent, nor as profitable, as that application area is today.

Chances are that your city, if it's large enough to show up on the map, has one or maybe even several electronic bulletin boards located in or near it. To list all of these would be an impossible task, or at least a project beyond the scope of this book. (If you want to check, the aforementioned *Computer Monthly* regularly contains a special section in which it publishes the connect numbers and access methods for hundreds of BBSs located in various areas of the country. In fact, those listings helped me research the admittedly cursory directory of BBSs I've included here.)

If the listings in my first two categories were incomplete, this one is positively bare bones. Literally thousands of BBSs have sprung up over the past decade, in virtually every part of the world. Given such a magnitude, the best I can hope for here is to hit the highlights. So I've provided, in alphabetical order by state, the following compendium of several representative BBSs. Each listing contains that board's geographic location, a brief description of some of the services it provides (in parentheses), its connect number, and specific information on that BBS's log-on protocols.

Note: Most BBSs provide free access, although some do request a minimal registration fee to gain complete access to all their features. Even if a given BBS is free, however, keep in mind that the phone company is still standing by, ready to charge you for any long-distance calls you place to contact a BBS. Just keep in mind that, like commercial information services, BBSs come and go.

Alaska

Anchorage Northern Lights BBS (SIGs, games, shareware, and public-domain software for Amiga users), (907) 337-4136, 300/1200/2400 baud, N-8-1.

Fairbanks Help Yourself BBS (concentrates on IBM-PC conferences, file areas, and shareware products, but includes message areas for all system brands), (907) 451-0887, 1200/2400 baud, N-8-1.

Alabama

Mobile Computer Confabulation (6 conferences, 18 doors, messages, large selection of IBM software), (205) 344-7606, 1200/2400/9600-baud, N-8-1.

Huntsville Publisher's Paradise (specializes on issues relating to desktop publishing), (205) 882-6886, 1200/2400/9600 baud, N-8-1.

Arkansas

Fort Smith Data Plus BBS (specializes in IBM and clone systems, file transfers, games, conferences), (501) 484-7821, 300/1200 baud, N-8-1.

Little Rock CAPCUA BBS (support board for Central Arkansas PC User Association), (501) 758-6045, 300/1200/2400 baud, N-8-1.

Arizona

Phoenix JCCS BBS Node 1 (technology, science, current events), (602) 582-3643, 300/1200/2400 baud, N-8-1.

Tucson Old Pueblo BBS (MS-DOS systems), (602) 744-2314, 300/1200/2400 baud, N-8-1.

California

San Jose SERVU (MS-DOS, Macintosh, CP/M80, CP/M86), (408) 238-9621, 300/1200/2400 baud, N-8-1.

San Francisco Abacus RBBS (dedicated to Atari users), (415) 587-8062, 300/1200/2400 baud, N-8-1.

Los Angeles BBS West (Supported by UCLA PC Users Group), (213) 473-2889, 300/1200/2400 baud, N-8-1.

Colorado

Denver MICRO BBS (downloadable shareware and misc. utilities for MS-DOS and CP/M systems), (303) 752-2943, 300/1200/2400 baud, N-8-1.

Colorado Springs Villa Hub (dedicated to TI99/4A and IBM computers), (719) 574-2567, 300/1200/2400 baud, N-8-1.

Connecticut

West Hartford University of Hartford BBS (Apple & IBM), (203) 242-4738, 300/1200/2400 baud, N-8-1.

Brookfield Wildside BBS (Echomail, DOS utilities, and programming information), (203) 775-4627, 300/1200/2400 baud, N-8-1.

Delaware

Newark FATHER Board (over 14,000 public-domain files, also provides access to PC-SIG), (302) 737-6041, 300 to 19,200 baud, N-8-1

Florida

Miami Beach Sunshine Online Service (Macintosh, MS-DOS, Apple, Commodore, Amiga), (305) 378-6828, 300/1200/2400 baud, N-8-1.

Ft. Meyer SWACKS BBS (sponsored by Southwest Florida Apple User's Group), (813) 334-4259, 300/1200 baud, N-8-1.

Jacksonville Galactic Infinitum BBS (all computer types), (904) 346-3087, 300/1200 baud, N-8-1.

Georgia

Atlanta IHBC BBS (information on running a home-based business) (404) 482-5753, 1200/2400 baud, N-8-1.

Savannah Savannah BBS (Supported by the Savannah PC Users Group), (912) 920-2006, 300 to 14,400 baud, N-8-1.

Hawaii

Maui Royal Hawaiian S/W BBS (group conferences and on-line entertainment), (808) 857-5651, 1200/2400/4800 baud, N-8-1.

Illinois

Cahokin Prolin BBS (aviation news, data libraries), (618) 337-0162, 300/1200/2400 baud, N-8-1.

Decatur U.S.S. Neversail (sponsored by the Commodore Computer Club), (217) 877-8726, 1200 baud, N-8-1.

Indiana

Indianapolis PC-DEN (MS-DOS systems, user help, games, conferences), (317) 862-5966, 300/1200/2400 baud, N-8-1.

Bloomington Indiana On-Line (Accessible through Indiana Bell Public Data Network), (812) 332-7227, 300/1200/2400/9600 baud, N-8-1.

Iowa

Cedar Rapids Racing Greyhounds (for anyone interested in the greyhound industry), (319) 362-7420, 300/1200/2400 baud, N-8-1.

Kansas

Wichita Presipus BBS (all computer systems), (316) 689-8771, 300/1200/2400 baud, N-8-1.

Kansas City KCKCC (MS-DOS systems, over 300 files), (913) 334-5511, 300/1200 baud, N-8-1.

Kentucky

Louisville DISK (multi-line UNIX system offering e-mail, multiplayer on-line games, and more), (502) 968-5401, 300/1200 baud, N-8-1.

Lexington Adventureland II (adventure games for most popular systems), (606) 273-6252, 1200/2400 baud, N-8-1.

Louisiana

Natchitoches Genesis BBS I (MS-DOS board with educational conference areas), (318) 352-8311, 300/1200/2400 baud, N-8-1.

New Orleans NOPCC RBBS (MS-DOS system sponsored by New Orleans PC Users Group and a FidoNet Node) (504) 436-1825, 300/1200/2400 baud, N-8-1.

Maine

Brunswick Baker Street BBS (Amiga user's board), (207) 725-1417, 1200/2400 baud, N-8-1.

Maryland

Baltimore Microline (private e-mail and over 20 subboards), (301) 922-3843, 1200/2400 baud, N-8-1.

Massachusetts

Westfield PVCC User Group BBS (sponsored by Pioneer Valley Computer Club, offers SIG doors to other systems), (413) 568-4466, 300/1200/2400 baud, N-8-1.

Plymouth Auto Exec BBS (specializes in database and programming issues), (508) 833-0508, 300 to 38,400 baud, N-8-1.

Michigan

Detroit Anchor BBS (sponsored by Michigan Commodore Users Group, Inc.), (313) 293-7340, 300/1200/2400 baud, N-8-1.

East Lansing The Program Exchange (large UNIX-like software library and message base), (517) 332-0472, 300/1200/2400 baud, N-8-1.

Minnesota

Minneapolis TCOS9UG BBS (sponsored by Twin Cities OS9 Users Group), (612) 780-8936, 300/1200/2400 baud, N-8-1.

Mississippi

Jackson MCS Micro Net (sponsored by Mississippi Computer Society), (601) 992-4023, 300/1200/2400/9600 baud, N-8-1.

Missouri

St. Louis SLACC STACK (e-mail, SIGs, shareware, and public-domain software), (314) 367-1903, 300/1200/2400 baud, N-8-1.

Kansas City COMP-SPEC (public-domain files for MS-DOS systems and compatibles), (913) 236-6530, 300/1200/2400 baud, N-8-1.

Montana

Dillon Montana Gold West BBS (Montana tourist information and electronic mail), (406) 683-6285, 300/1200/2400 baud, N-8-1.

North Carolina

Ashville Blue Ridge BBS (for IBM and Amiga users), (704) 258-9017, 300/1200 baud, N-8-1.

North Dakota

Fargo RBBS-PC of Fargo, Inc. (sponsored by Fargo IBM-PC Users Group; concentrates on law issues, Lotus, programming), (701) 293-5973, 300/1200/2400/9600 baud, N-8-1.

Nebraska

Omaha G.O.C.U.G. BBS (sponsored by Greater Omaha Commodore Users Group), 300/1200 baud, N-8-1.

Nevada

Las Vegas Bytes BBS (sponsored by Las Vegas PC Users Group), (702) 644-1157, 300/1200/2400 baud, N-8-1.

New Jersey

West Orange SportsBoard BBS (technical and sports-related local and national conferences), (201) 731-9425, 1200/2400/9600 baud, N-8-1.

New Hampshire

Bradford Inter-active Micro BBS (downloadable files for IBM systems), (603) 938-5230, 300/1200/2400 baud, N-8-1.

New York

New York E.R.I.C. (technology, programming, buy/sell), (212) 420-9380, 300/1200/2400 baud, N-8-1.

New York Wise Byte BBS (programming, UNIX, Novel Networks, and more), (202) 962-1946, 1200/2400/9600 baud, N-8-1.

Ohio

Cleveland PC-Ohio BBS (sponsored by Greater Cleveland PC Users Group), (216) 381-3320, 1200/2400/9600 baud, N-8-1.

Dayton The Annex (games, doors, adult areas, callback connection), (513) 274-0821, 300/1200/2400 baud, N-8-1.

Oregon

Portland Portland PCC BBS (sponsored by Portland PC Users Group), (503) 226-4142, 300/1200 baud, N-8-1.

Pennsylvania

Philadelphia Chat Attack! (15-line multi-user chat and game board), (215) 887-6600, 300/1200/2400 baud, N-8-1.

Harrisburg BREC BBS (FidoNet node and FOG member), (717) 657-8699, 300/1200/2400/9600 baud, N-8-1.

South Carolina

Hilton Head Island Palmetto Software Exchange (technical games, utilities, messages, etc.) (803) 671-9239, 1200/2400 baud, N-8-1.

Tennessee

Nashville TANSTAAFL BBS (religion, philosophy, programming) (615) 360-8419, 300/1200/2400 baud, N-8-1.

Virginia

Fairfax REMJEM On-Line BBS (sponsored by the Washington Area Commodore Users Group), (703) 503-9410, 300/1200/2400 baud, N-8-1.

Texas

Dallas Longhorn II BBS (sponsored by Dallas TI Home Computer Group), (214) 240-4979, 300/1200/2400 baud, N-8-1.

Washington

Seattle TIPS BBS (sponsored by Puget Sound 99ers), (206) 784-4142, 300/1200 baud, N-8-1.

Wyoming

Casper The Grasshopper Trap (sponsored by Casper Commodore Users Group), (307) 577-7400, 300/1200 baud, N-8-1.

Summary

There you have it, a fairly representative cross-section of commercial services, electronic mail companies, and local BBSs for you to experiment with once you graduate past remote computing and start using pcANYWHERE for more traditional on-line activities. And I've only scratched the surface, especially where independent bulletin boards are concerned. Almost any BBS you call will offer some sort of compiled list, usually in the form of a downloadable disk file, containing the names and access numbers for other BBS systems—both in its immediate area and in distant parts of the land.

After you've gotten your feet wet with the services listed here, you might decide to take the plunge and dive into InterNet, a truly advanced on-line environment. If so, check out the next and final appendix, written by my good friend Ralph Roberts. It describes this incredibly ubiquitous electronic gateway to literally hundreds of computers. If telecommunications provides you with a window to the world, as I've proposed several times throughout this book, InterNet is a wall of glass.

Ending with this little metaphor, I'll turn the actual discussion of InterNet over to Ralph. I know you're in good hands. As we say here in Indiana, he's good folk.

B
APPENDIX

Internet, Usenet, and other universes

by fellow writer and good friend, Ralph Roberts

Can you name the largest computer network in the world? You have over 5,000 chances of getting the right answer (or at least *a* right answer). It's Internet, of course, an often anarchic collection of over 5,000 local-area networks (LANs) located all over the world, all tied together by several wide-area networks (WANs) or, in Internet parlance, "backbones."

No one knows for sure, but some experts claim that Internet has as many as 10 million users. Hundreds of colleges and universities worldwide are connected to the network, as are government agencies of many countries, and all sorts of commercial companies. Because the network ties together so many researchers, it's also one of the most important computer networks in the world, in addition to being the largest. Internet is what the worm launched by Robert Morris, Jr. in November of 1988 essentially shut down for a night by overloading over 6,000 very important governmental and academic research computers. This caused a good deal of concern, a lot of attention by the mass media, and tightening of security procedures, but Internet remains open and essentially free to anyone who can obtain access. (Although Morris' worm wasn't really a computer virus, it was this attack that brought computer viruses to the attention of the public and made them a household term.)

How Internet started

Internet is perhaps the only military project in history that got totally out of hand and turned out to enhance the greater good. It was started by the U.S. Defense Department's Advanced Research Project Administration in 1973 as an experiment in interconnecting regional networks—using the names ARPANET for the more general portion, and MILNET for the military subnet. Other government agencies joined their computers to it—including NASA and the Department of Energy. Because a lot of research was done at universities, school systems began appearing on the network. The National Science Foundation (NSF), another U.S. government agency, also came into the picture, funding high-speed connections between research centers here and abroad.

Through the 1980s, Internet grew like crazy, spreading across oceans and losing its American governmental control. NSF still funds some of the Internet "backbones," but it's safe to say now that no one owns or controls the network. It simply exists—there for the use of anyone who accesses it.

Who pays for all this? Each institution, agency, or company pays for its own equipment and for any lines necessary to connect to Internet. NSF and similar agencies in various countries pay for the backbones to connect the 5,000 and more subnets together. There are few restrictions not to abuse the net, but there's a nebulous code of conduct known as *netiquette*, or net etiquette, that spells out a lot of what's acceptable and what isn't. Blatant commercial usage, for example, is frowned upon. A phenomenon known as *flames* will let you know when you cross the line. You'll get messages from a myriad of people in varying degrees of impoliteness letting you know exactly what you did wrong. Outspokenness and shyness are certainly not rare traits on Internet, that's for sure!

Accessing Internet

Access to Internet is, in one form or another, easy to obtain. If your school or company is on Internet and you have an account on their computer system, you might already have access or can get on by simply asking your system administrator to assign you Internet privileges. There are no charges to individuals—you get the world for free.

If you're not associated with an institution or firm whose computers are hooked up to the net, it's still easy enough to get on Internet. Many avenues are available to you. Perhaps one of the easiest is to find a local college or university that's willing to grant you an account on their computers in return for your contribution of service to that institution, or to Internet and the world community in general. Another option is to sign up with a commercial service offering an Internet gateway. One of the most reasonable in terms of cost is Netcom Online Communications Service in San Jose, California (call 1-408-554-UNIX voice for more information, or e-mail to bobr@netcom.com). For as low as $17.50 a month for a personal account, you can have full Internet access at up to 19,200 baud.

Navigating Internet

By using a modem and your personal computer, you can dial into a larger computer, then go out on Internet to almost anywhere in the world. Most computers on Internet are Unix based, so a minimal knowledge of that operating system will help you to navigate along the network from computer to computer, from country to country. The two main commands you'll need to use are telnet and ftp (file transfer protocol).

Telnet gives you full access to any computer for which you know the Internet address (thousands are published on the net and easily found). However, you must have a valid username and password to log on, or know a public account. For many universities, the username library will let anyone access the university library's card catalog. So if you have an account on the computer, you can access it from anywhere in the universe via Internet and operate just like you were using a terminal in the same building as the computer. However, it's unlikely that even Jack Nimersheim will wind up with a valid account on the many thousands of computers worldwide, and even less likely that he could remember all the associated usernames and passwords.

Because Internet exists primarily to encourage and facilitate the free exchange of information, most computer systems will allow you access even if you don't have an account with the ftp command. To access a computer in another city, state, or country that's on Internet, type ftp *internet_address*. Once a connection is indicated, type the word anonymous and, if asked, your username on the system that ties you into the net. If the computer allows public access, you can now use Unix commands such as ls to list the current directory and cd *new_directory* to change directories.

Most systems with public access have a /pub directory with subdirectories off of that containing programs and data files that you can transfer to your home system (using the get command). Because Internet connections are very fast, you can "get" files megabytes in length in only a couple of minutes or so. Alas, if you've dialed into your Internet gateway at 2400 baud, it might take you a couple of hours to get those same files across town and into your personal computer. Here's how Internet works.

Suppose, for example, that you live in Chicago but you want to see what books the Australian National University Library has on kangaroos. You connect to your local system on Internet and type the command telnet library.anu.edu.au (telnet is the the Unix command described above and the rest is the Internet address of the Australian National University Library). After being successfully connected, you'll see the screen shown in Fig. B-1.

Once you have the opening screen, follow the instructions and submit your search for the word *kangaroo*, but prepare yourself for a shock. The Australian National University Library only has 18 references with that subject keyword, and most of those refer to Kangaroo Island rather than the animal.

Obviously, another search term is required, so try the plural version instead and g'day, mate. On the other hand, the subject of kangaroos might not have you "hopping" with excitement. There are many thousands of computers at leading

```
System V.R2.4 UNIX (madmax)
+--------------------------------------------------------------+
| ==========================================================  |
|  Welcome to the Australian National University Library       |
| ==========================================================  |
|                                                              |
|  How to Login to our Online Public Access Catalog (OPAC):    |
|  ----------------------------------------------------------  |
|              * Terminal Emulation: VT100 required. *         |
|                                                              |
|                Type "library" at the login prompt.           |
|                                                              |
|  How to Exit:  "Ctrl ]" will take you back to your machine   |
|  -----------                      <or>                       |
|                Use the "EX" option from the top menu screen. |
|                                                              |
|Problems can be sent via e-mail to "helpdesk@library.anu.edu.au"|
|                          or phone (06) 249-2009              |
+--------------------------------------------------------------+
```

B-1 Initial screen for the on-line database of the Australian National University Library, which you can connect to through Internet.

universities and other research facilities all over the world hooked to Internet. Most of these allow some sort of public access, meaning there are literally millions of programs and data files that you can download into your own computer.

Archie

Let's say you have a personal computer running MS-DOS and Windows, and you want to find Windows-related programs useful to your home system. You could spend a century or so visiting a thousand huge computer systems, both here and abroad. A far better solution is to call your buddy "archie" up at McGill University in Montreal, Quebec. Archie is a computer connected to Internet that keeps track of programs and other files on those thousands of computers across the world. You tell archie the subject you want (online help is available), and archie will give you a list of the machines having matching files, their Internet addresses, and which directory you'll find the files in. Archie will even electronically mail you the results of its search (which can be very long for common terms).

To talk with archie, type telnet quiche.cs.mcgill.ca. Some of the best archives in the world for MS-DOS programs, by the way, are at wuarchive.wustl.edu (Washington University in St. Louis), garbo.uwasa.fi in Finland, wsmr-simtel20.army.mil at White Sands Proving Grounds in New Mexico (a military computer, but one with a large public access area), and msdos.archive.umich.edu at the University of Michigan. Old archie can be your best friend on Internet. Otherwise, you might never find the files you want.

Usenet

Usenet is not Internet, but parts of it overlap, everyone gets the same news, and e-mail can be exchanged between users on both networks. Usenet is a collection of computers, also worldwide, that exchange large packets of information consisting of electronic mail and that most wonderful of institutions (on both networks), the News.

The major difference between the two networks is that Internet is an interactive network, i.e., you actually log onto and operate the remote computer. On Usenet, information is sent from computer to computer, showing up on your Usenet system to be read or used. There's no realtime connection where you can do directory listings on the remote computer, or run its programs as you can on Internet.

Usenet was started in 1979 by two graduate students at Duke University in North Carolina (Tom Truscott and Jim Ellis). The first connection was to their archrival, the University of North Carolina in nearby Chapel Hill. From there, Usenet spread out to cover the world. Access to either network is worthwhile. If you have your own home Unix system, a Usenet connection so that you get the News is a very nice thing to have.

Usenet can be accessed over regular phone lines, so you can afford to be on Usenet from home. Very few people, however, will be able to afford a high-speed Internet connection, costing thousands of dollars a month. So your best bet is a modem connection to your local school or to a service like Netcom.

The News

Ah, now for the real fun part, the News. The News, which is on both networks, is the world's largest (by far) electronic bulletin board. There are literally hundreds of News groups on every topic under the sun, including Suns (a leading manufacturer of workstations—you'll often encounter them as computers on the net). The News has been described as "the most fun you can have with a terminal" and even "the most fun you can have with your clothes on."

You can communicate with people of similar or even wildly dissimilar interests, and exchange information with the world. Be warned, however, that the News is not only exceptionally useful, it's also addictive. Few of us on Internet or Usenet could survive without our wonderful daily fix of News.

Electronic mail

You'll make a lot of friends out there in the electronic universe. Both Internet and Usenet give you the excellent medium of electronic mail to keep in touch. Sending a message to a friend in Denmark isn't only free, but it sometimes arrives only moments after you send it (depending on how the network routes it).

Conclusion

Alas, space here precludes a fuller discussion of Internet and Usenet—I would need a complete book to do the subject justice. However, this should introduce you to the electronic universe out there, and whet your appetite to travel in the wondrous realms of cyberspace along the silver strands of Internet and Usenet.

Glossary

analog signals The type of "tones" normally transmitted over standard telephone lines. A modem converts the digital signals generated by your PC into analog tones during an on-line session.

ASCII Acronym for American Standard Code for Information Interchange. ASCII is the most "vanilla" format available for storing text files. Basically, an ASCII file contains a series of numbers, each of which corresponds to an access code.

asynchronous A communication method in which information exchange is coordinated by separating individual characters with start and stop bits, rather than using a prearranged, uniform time period to alternate data transmission between the two modems. Most PC-to-PC modem communications are asynchronous connections.

AT command set A series of commands used to manage telecommunications, first popularized by Hayes modems. This command set derives its name from the fact that you precede its commands with the letters AT to get the attention of your modem. The AT command set has evolved into a standard for PC telecommunications.

auto answer The condition where your modem has been instructed to automatically answer any incoming calls and then generate a carrier tone to establish a connection with a remote computer.

automated procedure A series of preprogrammed steps used to automate DOS command execution or file transfer on the host and remote computers during a pcANYWHERE session.

baud rate A measurement of the number of times a modem changes the frequency of its carrier signal. Though expressed as bits per second (bps), a modem's baud rate is not necessarily a true measurement of how fast data actually is transmitted during a communications session.

BBS Abbreviation for Bulletin Board System. Appendix A of this book contains a partial listing of BBS systems located throughout the United States.

binary notation A notation system that uses only two numbers. Computers rely on binary notation to indicate the status of their electronic switches, with 0 and 1 used to represent off or on, respectively.

bit Contraction for binary digit, the smallest information unit that can be identified by a computer.

block A cluster of data transmitted as a single unit during a telecommunications session. Most file-transfer protocols break large files down into discrete blocks before transmission.

bps Abbreviation for bits per second. This is the actual number of bits (binary digits) transferred every second during an on-line session. The bps measurement can differ from baud rate because it often takes multiple state changes (baud) to transmit a single bit.

buffer A portion of RAM set aside to temporarily store data. Many communication programs use a data buffer to manage incoming data during an on-line session.

byte The primary unit of measurement for computer memory. As a rule, it requires one byte, either 7 or 8 bits in length, to represent each character transmitted during a modem session.

caller entry An individual item in a list of entries pcANYWHERE/Host uses to record important information about the remote systems that will be calling it.

carrier detect The process used by a modem to indicate the presence of a signal.

carrier signal A tone generated over a telephone line that can initiate data exchange between two computers involved in a modem connection.

chat mode Used during a pcANYWHERE session to allow users at the host and remote computers to communicate with one another over the remote connection. When invoked, chat mode suspends remote operations and displays a window on both users' screens, which they then use to type comments.

communications port Also called a COM port, this is actually a serial port installed on your PC. You must specify the COM to which your modem is attached for your communications software to work properly.

communication parameters The rules two computers use to communicate with one another during a modem connection. Protocols include items such as baud rate, parity, start bits, stop bits, data rate, file-transfer methods, and the like.

communication software Any program that allows your PC to use a modem and standard telephone line to contact a remote system.

cps Abbreviation for characters per second, or the number of characters transferred each second during an on-line session. Though this differs from the bps measurement, the two are related. If a computer requires eight bits to represent each character, then the normal cps speed for an on-line session involving that PC could be approximately calculated by dividing its bps rating by 8.

CPU The central processing unit, the computer chip that is the "brains" behind your PC. Most MS-DOS computers use one of three CPU types: 8088/86 (PC/XT), 80286 (PC-AT), and 80386, all of which were designed by Intel Corporation. Recently, Intel introduced the 80486 CPU.

data rate The speed at which data is moving across a telephone connection. Data rate is generally measured in bps, or bits per second.

data word An organized group of bits. Individual data words are organized and separated using start bits and stop bits.

default settings The normal configuration of a piece of PC hardware or software when it's first turned on. As a rule, communication programs allow you to specify the default settings for the majority of your on-line activities.

digital signal The types of signals generated by your PC. Generally, a digital signal is represented by either an on or off state, a 0 or 1, or some such notation. A modem converts the analog tones coming over a phone line into the digital signals required by your PC during an on-line session.

display adapter The type of monitor and video board you have attached to your PC.

download The process of transferring files and/or information from a remote system to your PC.

duplex A communications parameter used to indicate how characters entered at the keyboard are transmitted to your monitor. In full-duplex mode, characters are visible only on the local monitor after they're sent to the remote system, which retransmits them back over the line for display. In half-duplex mode, characters are sent directly to your monitor without first making the journey to and from the remote system.

e-mail Abbreviation for electronic mail. The process of transferring messages (usually ASCII text files) over a modem and telephone line.

emulation The process of configuring your PC to resemble a specific terminal type. Many communication programs, for example, allow you to emulate a DEC VT100 terminal to enable PC-to-mainframe communications.

external modem A stand-alone modem that isn't installed inside your computer. You must generally connect an external modem to your PC through a serial port.

full duplex A condition where letters and commands you enter on-line are first sent to the remote system, then echoed back to your display monitor. (If you can't see what you type on your monitor, you should change the duplex setting for your communications software to half duplex.)

half duplex A condition where letters and commands you enter on-line are immediately sent to the display monitor by your PC. (If everything you type during an on-line session appears twice on your monitor, e.g., yyoouu ssee tthhiiss, then you probably need to change the duplex setting for your communications software to full.)

hardware configuration A collection of settings that pcANYWHERE uses to control communications with another computer.

host entry An individual item in a list of entries pcANYWHERE/Remote uses to record important information about the host systems it will be calling.

host system A computer (either a PC or mainframe) set up to accept calls from another system. This is the computer running pcANYWHERE/Host during a pcANYWHERE session.

hot key A special key or combination of keys used to call up a pcANYWHERE menu during a remote-computing session.

internal modem A modem you install directly into an expansion slot inside your PC.

kilobyte A unit of measurement generally used to specify file size. One kilobyte, or K, equals 1024 bytes.

local system A relative term used to identify the computer you're entering commands into during an on-line session. *See* remote system.

log A record of all successful connections established by or with a system running pcANYWHERE. Logs can be maintained on either the host system or remote system.

master password An optional password you can assign to your pcANYWHERE environment. If a master password is defined, pcANYWHERE will ask you to enter it before it will allow you to modify program settings, host configurations, caller entries, and the like.

modem The device used to convert analog tones to digital signals, and vice versa, during an on-line session. Modem is a contraction for *mo*dulate/*dem*odulate, the technical term for this process.

modulate To change or modify some aspect of a carrier signal in a prearranged manner in order to transmit information over a modem connection.

non-pcANYWHERE host A host system that isn't running pcANYWHERE/Host. Commercial services and local BBSs would be two examples of a non-pcANYWHERE host.

off-line The condition of not being connected to a remote system.

on-line The condition of being connected to a remote system.

packet-switching network An electronic switchboard of sorts that can be used to contact a number of on-line systems. The two most popular packet-switching networks for PC communications are SprintNet and Tymnet.

parameters *See* communication parameters.

parity An error-detection technique used to verify the integrity of data transferred over a host/remote connection.

procedures *See* automated procedures.

protocol The rules and line settings two computers use to initiate and then verify file transfers and data transmission. As a rule, PC communications programs support a number of popular file-transfer protocols.

RAM Random-access memory, the dynamic memory on your PC that's used to load and execute application programs.

real-time conference An on-line session during which several users communicate with one another over a modem link. Most commercial information services sponsor regular real-time conferences in a number of areas of special interest.

remote system In general telecommunications, a relative term used to identify the system you're using to communicate with the host computer during an on-line session. In a pcANYWHERE session, the system running the pcANYWHERE/Remote software. *See* local system.

ROM Read-only memory, nonvolatile memory used to permanently store information in your PC. For example, your computer's initial startup procedures are stored in ROM. The contents of ROM cannot be altered by the user.

script A series of preset sequences or activities that execute whenever that script is run. Scripts allow you to automate many on-line activities you would otherwise have to initiate manually.

search The process of locating particular information during on-line research. Most research systems allow you to use Boolean logic to extract specific records from their databases.

SIG Abbreviation for special-interest group. Generally, a special area within a commercial on-line service or BBS where individuals with shared interests meet. (On CompuServe, SIGs are more commonly referred to as forums.)

start bit An identifying bit transmitted during an asynchronous modem session. The start bit marks the beginning of a character.

stop bit A data bit that marks the end of a character during an asynchronous modem session.

sysop Abbreviation for system operator, the person generally responsible for maintaining a BBS or a specific SIG area within a BBS (or on-line commercial service).

telecommunications The process of using your PC, a modem, and a special program to communicate with another computer over standard telephone lines.

upload The process of transferring files and/or information from your PC to a remote system.

[illegible] files. See automated procedures.

protocol: The rules and procedures two computers use to govern their file transfers and data transmission. As a rule, PC communications programs support a number of popular file-transfer protocols.

RAM: Random-access memory, the dynamic memory in your PC that's used to load and execute application programs.

real-time conference: An on-line session during which several users communicate with one another over a modem link. Most commercial information services support [illegible] real-time conferences [illegible] special interest [illegible]

remote system: In general data communications, a relative term used to identify the system [illegible] the host computer during an on-line session. [illegible] REMOTE [illegible]

ROM: Read-only memory [illegible]

script: [illegible] instructions show [illegible]

search: The process of locating [illegible] records or information [illegible]

SIG: [illegible]

start bit: [illegible] the start of [illegible]

stop bit: A data bit that marks the end of [illegible] asynchronous modem session.

SYSOP: Abbreviation for system operator, the person generally responsible for maintaining a BBS or [illegible] within [illegible] on-line commercial service.

telecommunications: The process of using [illegible] a modem and [illegible] to communicate [illegible] standard telephone lines.

upload: The process of transferring [illegible] from your PC to a remote system.

Index

B

C

D

I

N

O

P

Q